DATA NEWBIE TO GURU

The Visual Guide to Learning SQL
in a Week
1st Edition

DR. CHANTEL A. PERRY

BIONIC BOOKS
DIVERSIFYING STEM EDUCATION

DATA NEWBIE TO GURU

THE VISUAL GUIDE TO LEARNING SQL IN A WEEK

Published by
theseen.ai llc
945 North Point Dr. #1292
Alpharetta, GA 30022

Notices

Knowledge and best practice in this field are constantly changing. As new research and experience broaden our understanding, changes in research methods or professional practices may become necessary.

Practitioners and researchers must always rely on their own experience and knowledge in evaluating and using any information or methods described herein. In using such information or methods they should be mindful of their own safety and the safety of others, including parties for whom they have a professional responsibility.

To the fullest extent of the law, neither the Publisher nor the authors, contributors, or editors assume any liability for any injury and/or damage to persons or property as a matter of products liability, negligence or otherwise, or from any use or operation of any methods, products, instructions, or ideas contained in the material herein.

Support & Bulk Orders
bionicbooks@theseen.ai

Training, Press and Media
hi@theseen.ai

ISBN: 978-1-7346836-2-2

Website:
www.dn2g.com

CREDITS

EDITOR
Dr. Cindy Hanson

TECHNICAL EDITOR
Carlotta Allen

PEER REVIEWERS
Dr. Paul Williams
Tobe Okeke
Tim Steward

COVER DESIGNER
Stas Snigur

DESIGN AND LAYOUT
Maureen Manalo

CODE TESTER
Maureen Manalo

DEDICATION

Thank you to all my loved ones who support me:

Chris, Lottery, Dr. Charles, Evelyn Jaunita, Levar, Bonnie, George, Geli

Samzee, Tenesha, Keisha, Candace, Jack, Dr. P

Thank you to all my teachers who motivate me:

Mr. Booten, Dr. Kevin Ryan, Dr. Mark Culp, Dr. Phillip Turk, Dr. Zondlo

Thanks to those who inspire me to do more for myself and community:

Coach Ryan, Dr. Brandeis Marshall, Chris, Tobe

OUR MISSION

ADVANCING EQUITY THROUGH DATA, TECH, & AI

BY PURCHASING THIS BOOK YOU HAVE BECOME A MEMBER OF TEAM #DIVERSIFYDATA
EVERY **100** BOOKS SOLD PROVIDES A HISTORICALLY DISENFRANCHISED GROUP WITH

10 BOOKS
A COURSE CURRICULUM
A PROFESSIONAL DEVELOPMENT SCHOLARSHIP

Additionally, **10%** of profits are donated to equity focused R&D and education programs
theseen.ai

GETTING STARTED

Thank you for purchasing this book. We hope this visual guide will assist you in accomplishing your SQL needs.

This book was written for:

- Beginners who desire to get started in computer programming, database management, or analytics
- Those who already have knowledge of databases, analytics, and programming, but would like to know more about how to utilize data

To use this book you will need:

- MySQL installed on your computer
- Data Guru Follow Along Files

Please visit mysqlw10.dn2g.com for guided installation instructions for Windows 10 computers and mysqlmac.dn2g.com for MacOCs devices. Download & run all the code for this book online at gitlab.dn2g.com

BOOK STRUCTURE

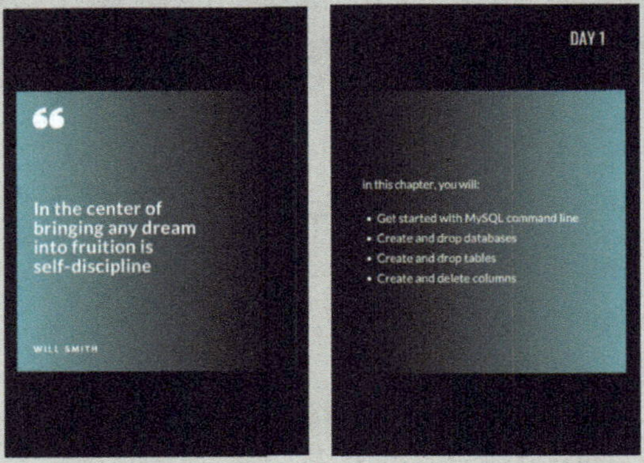

Topic Intro pages are displayed at the beginning of each lesson and briefly introduce the concepts covered in the following chapter.

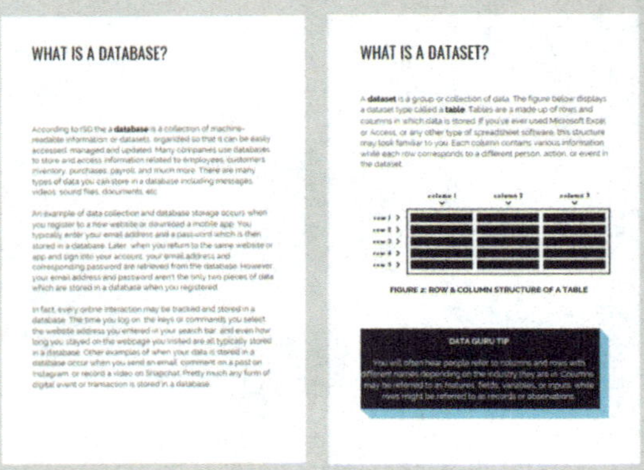

Concept pages provide explanations of the topic. These pages have a white background.

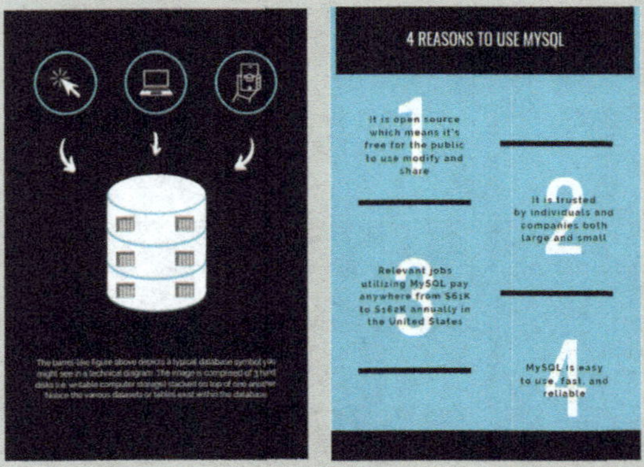

Diagrams and infographics have a black or teal background. They provide a visual illustration to support learning concepts.

BOOK STRUCTURE

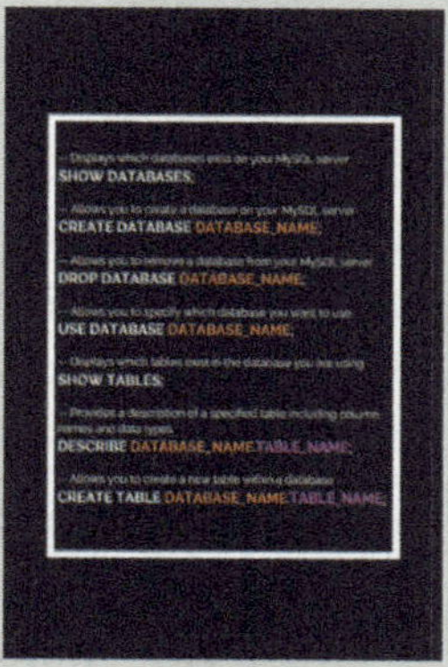

Syntax pages include code to help you understand SQL rules. These pages have a black background with a white box and embedded text

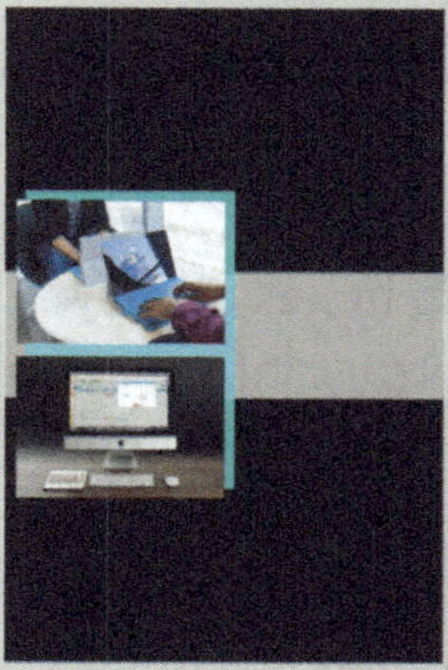

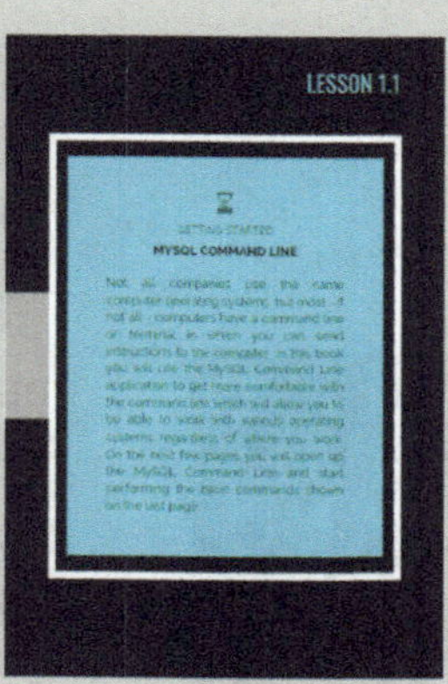

Lesson pages indicate a hands on lesson. They are black and the lesson descriptions are shown on a teal background.

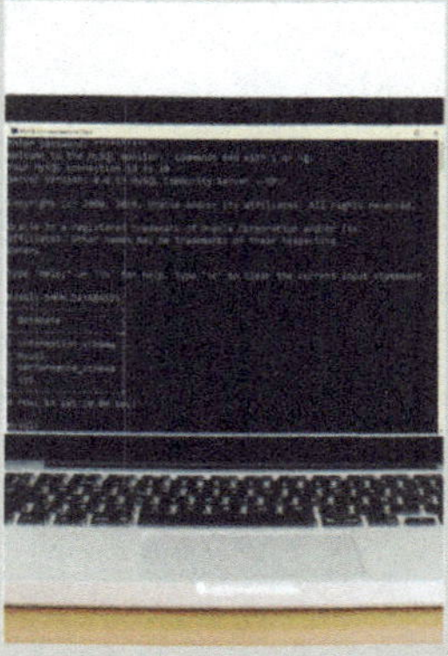

Exercise pages contain instructions with sample code illustrated on a white background. Examples of the code are depicted on a computer screen on the subsequent page.

BOOK STRUCTURE

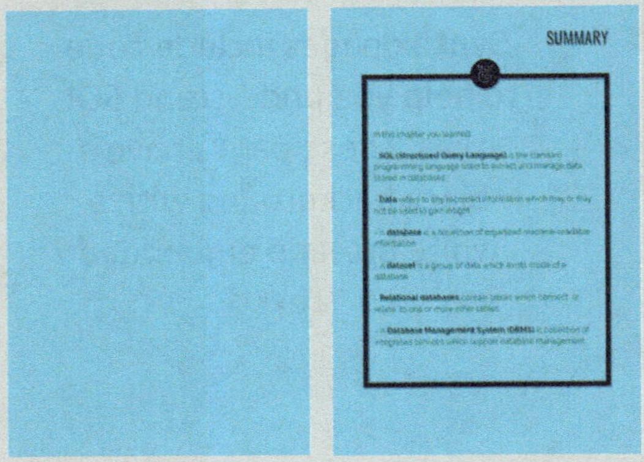

Summary pages are displayed on a teal background and are intended to reiterate key concepts discussed in the chapter.

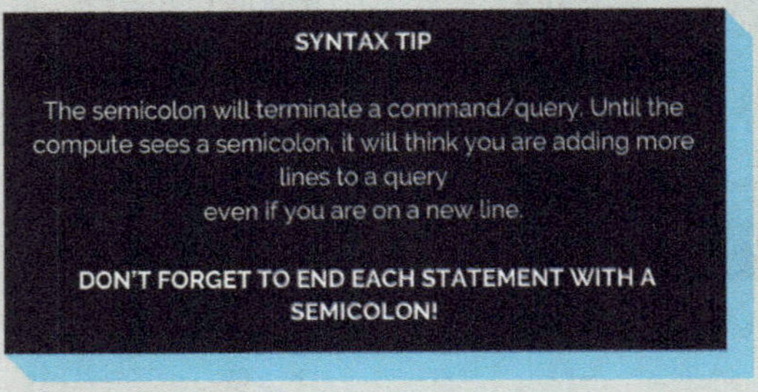

Tip boxes display important information for you to remember. They are labeled with the word "TIP"and appear in a black rectangle with a teal shadow.

Additional information in the form of clickable websites or hyperlinks will have an underline.

"

The world is now awash in data and we can see consumers in a lot clearer ways

MAX LEVCHIN
PAYPAL CO-FOUNDER

INTRODUCTION

In this chapter, you will learn:

- Basic concepts about what data entails and how databases work
- Key concepts in database management
- How data is collected from online interactions
- The importance of learning SQL
- Why MySQL is the software provider used throughout this book

WHAT EXACTLY IS DATA?

Gone are the days when data simply referred to statistics, facts, and calculations. The term **data** refers to any recorded information which may or may not be used to gain insight.

Globally ***2.5 quintillion bytes of data*** are created daily, with 90% of the world's data having been created within the past two years. This data is compiled from many sources: comments posted on social media sites, digital pictures and videos recorded on a cell phone, banking transactions listed on a banking statement, and even location coordinates displayed on a GPS.

In 2011, the McKinsey Global Institute reported that the production of data was increasing by a rate of 40% each year. By year 2020, Thompson Reuters News Analytics expected the total production of digital data to increase to approximately 35 zettabytes (equal to 35 sextillion bytes).

Due to the influx in data production, there is a growing demand for data gurus in the United States to fulfill needs in the tech industry such as data management, information security, and analytics. Unfortunately the U.S. is falling short in filling these positions, but that's where gurus like you and I come in 😎

DID YOU KNOW??

In 2018, more than 150,000 Data Science jobs were vacant and the demand is growing exponentially. There are plenty of well paying data based jobs and high-demand business opportunities in the field.

2.5 QUINTILLION BYTES OF DATA IS EQUIVALENT TO

1.25 billion 2gb USB drives

39,062,500 64gb iPhones

2.5 million 1tb PS4 game consoles

WHAT IS SQL?

SQL stands for **Structured Query Language** which is a programming language used to extract and manage data stored in databases (Note: we will cover databases more in depth in the next section). SQL is the industry standard for database language and has been adopted by both the American National Standards Institute (ANSI) and the International Organization for Standardization (ISO) [1]. SQL is the #1 programming language requested on Indeed for data professionals.

Learning SQL can help you effortlessly acquire employment and earn a higher salary as a data professional in the tech field. The table below lists careers, which utilize SQL, and the average annual United States salary for each position [2].

JOB TITLE	AVERAGE SALARY
DATA ANALYST	$69,291
BUSINESS INTELLIGENCE ANALYST	$80,430
SQL TESTER	$83,825
SQL DEVELOPER	$91,080
DATA ENGINEER	$114,887
DATA SCIENTIST	$120,495

1 Most technical definitions in this book are from the ISO site, which is managed by the organization that sets international standards across numerous industries and organizations. I highly suggest referring to their online browsing platform when you need to search for accurate definitions, symbols, and relevant publications https://www.iso.org/obp/ui/#home

2 https://www.dataquest.io/blog/why-sql-is-the-most-important-language-to-learn/

SQL CAREERS BY AVERAGE SALARY

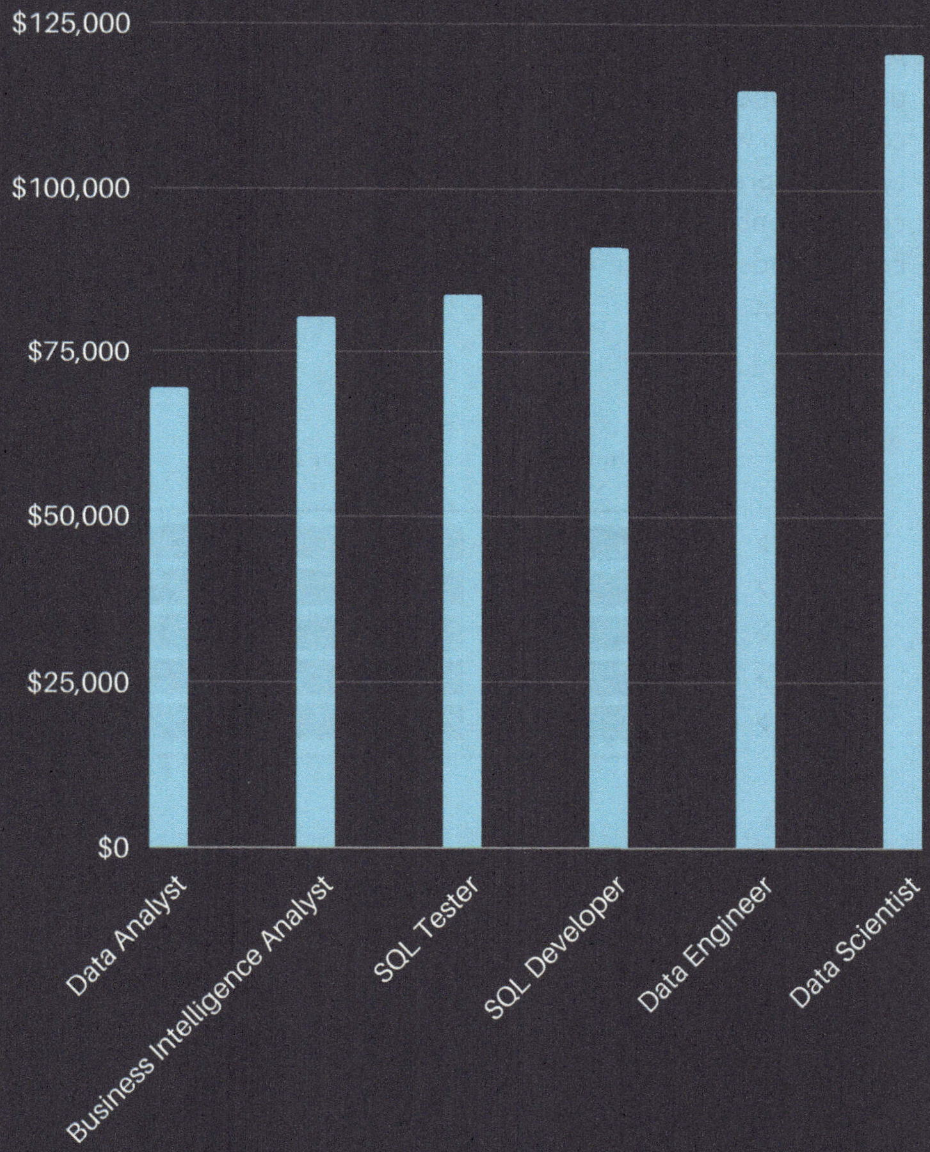

Average salary are based on searches done on Glassdoor and Salary.com websites on 11/26/2019. **Results are subject to change

WHAT IS A DATASET?

A **dataset** is a group or collection of data. The figure below displays a dataset type called a **table**. Tables are a made up of rows and columns in which data is stored. Microsoft Excel, Access, and other types of spreadsheet software are examples of table structures. Each column contains various information while each row typically corresponds to a different object, person, action, time, or event in the dataset.

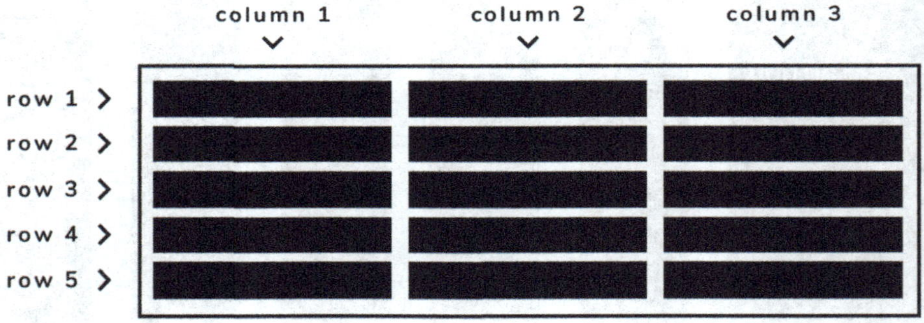

DATA GURU TIP

In different industries, you may hear others refer to columns and rows with different names. The term "column" may be interchanged with the *feature*, *field*, or *variable*, while "rows" might be referred to as *records* or *observations*.

TABLE DATASET EXAMPLE

STUDENTS		
StudentID	**FirstName**	**MobileNumber**
0000001	Mariella	(424) 444-2067
0000002	Cathrine	(768) 613-7373
0000003	Eleonora	(707) 816-5715
0000004	Alyce	(987) 846-5521
0000005	Noboku	(472) 377-6415

The **STUDENTS** table provides a snapshot of one dataset that you will use throughout this book. The students data stored in this table is similar to what you might see in a college or university database. Each **column** represents a different category of information: student identity number, first name, and mobile phone number. While each **row** contains unique data that corresponds to the individual specified in that row.

WHAT IS A DATABASE?

A **database** is a collection of machine-readable information, or datasets, organized so that it can be easily accessed, managed and updated. Many companies use databases to store and access information related to employees, customers, inventory, purchases, payroll, and much more. There are many types of data can be stored in a database including messages, videos, sound files, documents, etc.

An example of data collection and database storage occurs when a shopper creates an online account that retains their personal information in order to make consecutive purchases via an eCommerce website. The shopper enters their email address and password which is stored in a database. Later, when the shopper returns to the website to make another purchase, they are required to sign into their account. The email address and password stored in the database are compared to the sign-in information. When online shoppers create an account, their email address, password, and other registration data are not the only data elements stored in a database.

In fact, every online interaction may be tracked and stored in a database such as login time, buttons clicked, and website addresses. A few more examples of online interactions resulting in digital database storage include sending emails, connecting with other users on social media, commenting on restaurant review sites, and more. Always be cognizant of personal actions online because most digital events are stored in a company database.

DIGTIAL DATA STORAGE DIAGRAM

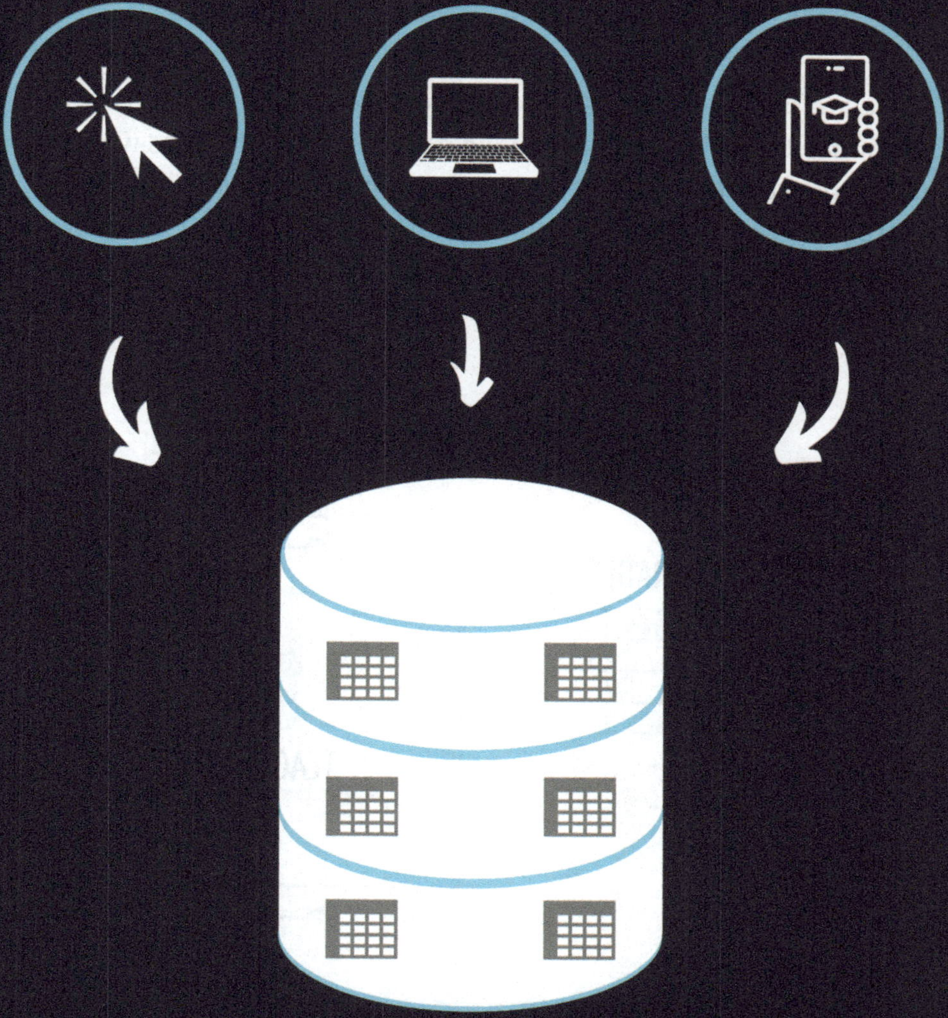

The barrel-like figure depicts a typical database symbol shown in standard technical diagrams. The figure is comprised of 3 hard disks (i.e. writable computer storage) stacked on top of one another. Notice various datasets or tables exist within the database.

RELATIONAL DATABASES

There are two major types of databases—relational and non-relational. In a **relational database** data is stored in tables where each table connects, or relates, to one or more tables. When tables relate they will have one or more rows in each table that uniquely correspond.

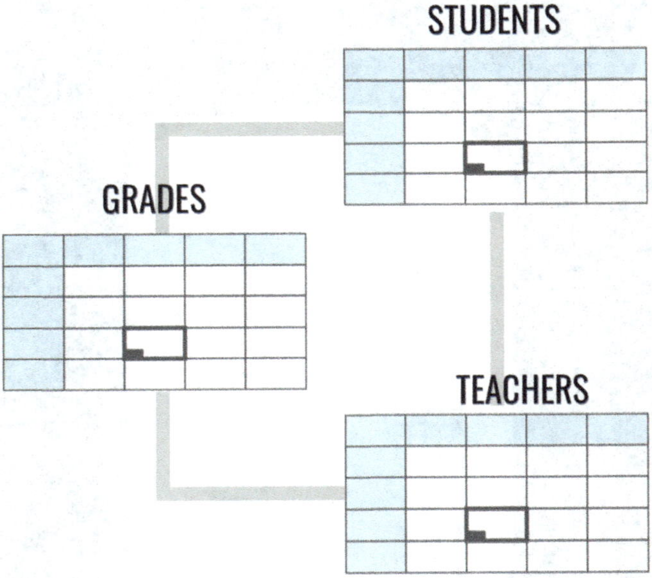

As stated, the tables have a relationship with one another meaning the data stored in each table is different, but they connect in some manner. See the example shown on the next page. The **STUDENTS** table contains general student enrollment information (i.e. names, phone numbers) and the **GRADES** table contains course grades for various classes.

EXAMPLE TABLES

STUDENTS

*StudentID	FirstName	MobileNumber
0000001	Mariella	(424) 444-2067
0000002	Cathrine	(768) 613-7373
0000003	Eleonora	(707) 816-5715
0000004	Alyce	(987) 846-5521
0000005	Noboku	(472) 377-6415

GRADES

StudentID	Course	Score
0000001	MATH 120	95
0000002	ENGR 385	67
0000004	ENGR 385	89
0000006	STAT 012	89
0000007	HIST 396	63

In a relational database table, **keys** or identifiers are used to relate one or more tables. Notice the column **StudentID** exists in both tables. Each student has a unique identity number which is used to connect their information in one table to the other. This type of unique identifier is called a **primary key.** The **primary key** is covered in detail on Day 6.

NoSQL DATABASES

Non-relational databases, also known as **NoSQL databases**, may or may not store data in a defined structure such as a table. Unlike relational databases, NoSQL databases are not required to have keys or relate to one another.

Don't fall into this common Data Newbie delimma
NoSQL should not be mistaken as a coding language.

NoSQL is a category of database types which stands for "**Not Only SQL**". This means that SQL can be used to manipulate NoSQL databases, but it's n*ot the only programming language* that can be used.

There are four types of NoSQL database: **key-value, document store, columnar, and graph-based**. Each of these database types is complex in nature and an entire guru book could be written for each of them. Since time is of the essence, this publication focuses on using SQL to manipulate relational databases. However, the next few pages briefly introduce the concept of non-relational database to provide a general understanding of how they work.

KEY-VALUE

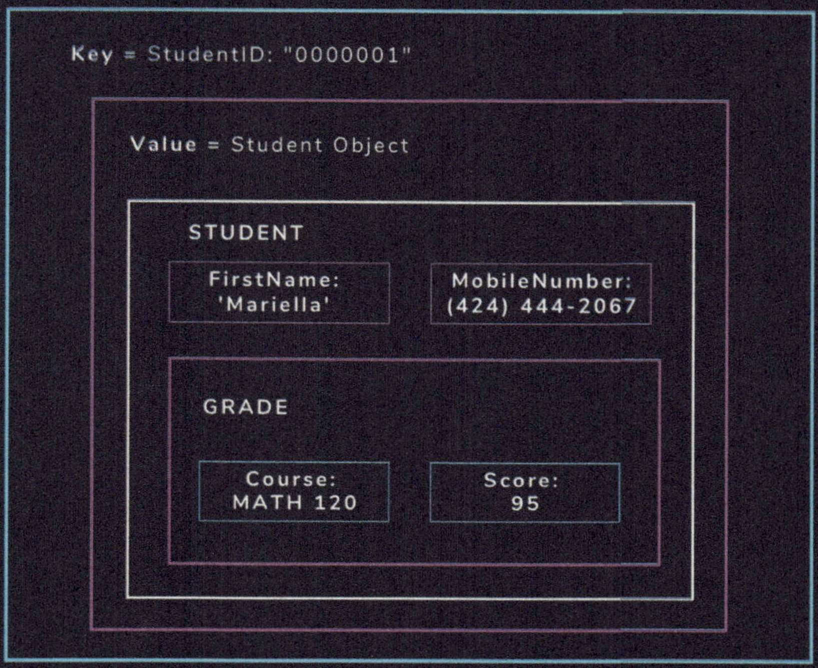

Key	Value
"0000001"	{STUDENT : {FirstName: 'Mariella'}; {MobileNumber: (424) 444-2067}; {GRADE : {Course: 'MATH 120'}; {Score: 95}
"0000001":STUDENT: FirstName	''Mariella'
"0000001":STUDENT: GRADE	{Course: 'MATH 120'}; {Score: 95}

Key-value databases store information in pairs; the parent object is called the **key** and the child object is called the **value**. This structure allows for storing data in a nested manner similar to folder directories on a computer.

DOCUMENT STORE

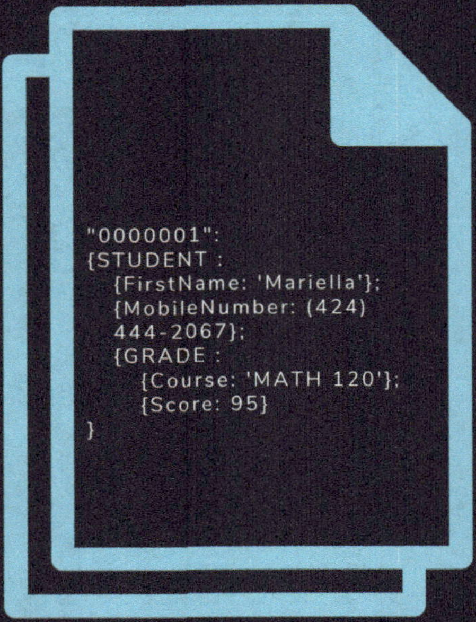

Document store databases are similar to key-value databases because both have a key-value structure. The difference is that data in document store is typically stored in an XML or JSON document.

COLUMN ID 01

STUDENTS
StudentID
0000001
0000002
0000003
0000004
0000005

COLUMN ID 02

STUDENTS
FirstName
Mariella
Cathrine
Eleonora
Alyce
Noboku

COLUMN ID 03

STUDENTS
MobileNumber
(424) 444-2067
(768) 613-7373
(707) 816-5715
(987) 846-5521
(472) 377-6415

Columnar databases are comparable to tables in a relational database except each column is segmented into a separate file and stored. Columnar databases are beneficial for searching through large tables because searching for data in a individual column is much faster than examining each row of data in a table.

GRAPH-BASED

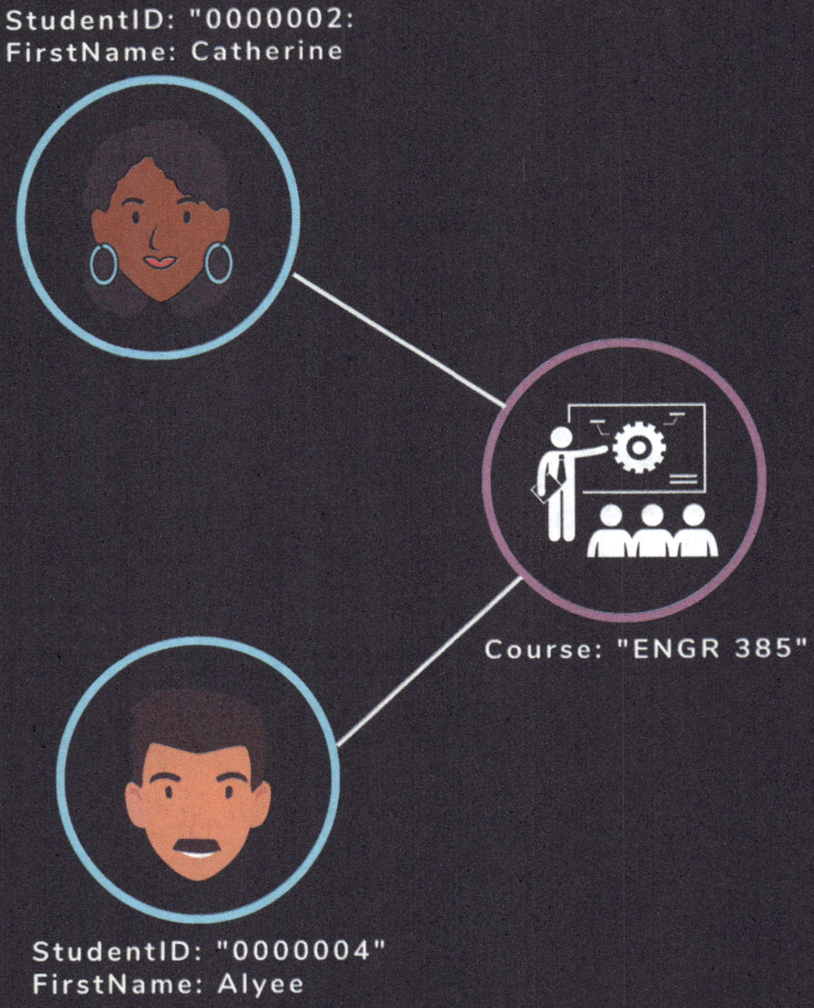

Graph-based databases use a network-like structure called a graph that is made up of nodes and edges. Each node contains information about an entity and the edges of the nodes connect one node to another. Graph databases are helpful with understanding relationships as illustrated above (i.e. Catherine & Alyee took "ENGR 385" together).

WHAT IS A DATABASE MANAGEMENT SYSTEM?

A **Database Management System (DBMS)** is a collection of integrated services which support database management and control the creation, use and maintenance of a database. A DBMS allows the user to create and store data in multiple databases and to control levels of access among different users.

You might be wondering...

"Why would I want to store data in different databases?"

Imagine running a mobile app development company and assume you've created a few apps. One for gaming, one for appointment booking, and one for fitness videos. If you continue using the same database to store information for your apps, there's a possibility that one of the following scenarios could occur:

- When the database goes down for maintenance or if it happens to crash, all apps will be unavailable causing customer service interruptions and possible monetary loss across the board;
- If a hacker breaches the database, data storage for all apps is at risk of being compromised;
- The database will expand faster resulting in the need to scale up sooner

These are just a few challenges businesses face when utilizing one database for multiple use cases. Some companies create different databases for each organization, department, or project depending on their needs.

SCENARIO: APP COMPANY DBMS

DO THIS

Create different databases when information exists for uniquely different purposes or there is no clear relationship

NOT THIS

When producing a lot of information, splitting it into different databases can help increase database security and reduce scalability issues.

WHAT IS MySQL?

MySQL is a software tool which allows users to quickly and easily create a **Relational Database Management System (RDBMS).** Other common RDBMS software development companies include Microsoft SQL Server, Oracle Database, and PostgreSQL.

In this book, we will use MySQL to create and manage databases. You may be wondering...

"Why are we using MySQL instead of another RDBMS software provider?"

Below you will recognize some of the top companies that use MySQL. and I've shared a few great reasons to learn SQL using MySQL.

4 REASONS TO USE MYSQL

1

It is open source which means it's free for the public to use, modify, and share

2

It is trusted by individuals and companies both large and small

3

Relevant jobs utilizing MySQL pay anywhere from $61K to $162K annually in the United States

4

MySQL is easy to use, fast, and reliable

HISTORY OF DATABASE STORAGE

Data storage methods have changed drastically over the years as technology has advanced. Let's briefly examine the evolution of the punch card. The **punch card** was invented by Joseph Marie Jacquard, a french weaver and merchant, in the early 1700s. He used the punch card to automate the textile weaving process through his invention the *Jacquard machine.*

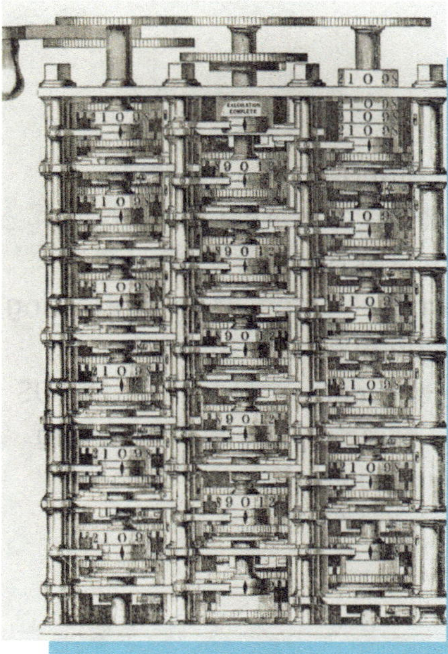

By the early 1800's, Charles Babbage - an English mathematician, philosopher, and inventor - adapted the punch card for use in programmable computing. He was credited for the invention of the first automatic digital computer and was deemed by some the "father of the computer".

HISTORY OF DATABASE STORAGE

Fig. 5. — Vue d'ensemble de la machine à compter (machine Hollerith). La presse P est représentée avec plus de détails dans le n° 2.

In the late 1800s, Herman Hollerith, an American businessman, inventor, and statistician, created a method to store data on punch cards. He developed a mechanical tabulator that performed statistical calculations using data from punch cards. As an interesting note, Hollerith's machine was commissioned by the United States government to assist in the data collection and processing of the US census. Soon after, Hollerith founded a company called Computing Tabulating & Recording (CTR), which subsequently become the iconically well-known International Business Machines (IBM) Corporation.

HISTORY OF DATABASE STORAGE

By the 1960s, punch cards were the standard medium for data input and storage. Initially, computers required the data on punch cards to be organized in a sequential format, usually with one line of code per card. Programmers loaded decks of coded cards into a card reader in order to input files or programs into the computer. Dropping the cards required re-sequencing of the deck.

Thankfully, technical innovations brought about the invention of the database and we no longer have the burden of coding and transporting large stacks of punch cards to the computer for data input, usage, and storage. Databases allow users to collect and store information digitally on a computer. Today, it's popular for businesses to use data centers containing database servers.

HISTORY OF DATABASE STORAGE

Over the course of time, different database systems have been tested. For example, IBM created a hierarchical database system called IMS DB that consisted of a network like structure used to develop and manage data. However, it was complex and eventually overshadowed by the invention of the relational database.

When Tedd Codd, an IBM employee, invented the Relational Database Management System (RDBMS), it was originally called System R. In order to interact with RDBMS, Don Chamberlin and Raymond Boyd developed Structure Query Language (SQL) which was commonly pronounced as SEQUEL. You can find the original introduction to the concept of SQL in a book titled"A Relational Model of Data for Large Shared Data Banks". Although SQL allowed users to create and manipulate data, it was not immediately released publicly after its inception.

Since the adoption of SQL as the standard database language many organizations have created their own versions. Though companies have modified versions of SQL to fit their needs, the basic syntax remains constant. You will notice the difference when viewing SQL rules from different RDBMS providers (i.e. MySQL vs. PostRegSQL).

While SQL remains the most popular database language, the world's Big Data problem has lead companies to utilize cloud storage, or to store and retrieve data from remote data centers. Cloud storage was invented by Dr. Joseph C. R. Licklider in the 1960s. Roughly 30 years later, AT&T released the first web based cloud service. Today we have many cloud storage providers such as Amazon AWS, Google Drive, DropBox, and iCloud among many others.

SUMMARY

In this chapter you learned:

- **Data** refers to any recorded information which may or may not be used to gain insight.

- A **database** is a collection of organized machine-readable information.

- A **dataset** is a group of data which exists inside of a database.

- **Relational databases** contain tables which connect, or relate, to one or more other tables.

- A **Database Management System (DBMS)** is collection of integrated services which support database management.

- **SQL (Structured Query Language)** is the standard programming language used to extract and manage data stored in databases.

- **MySQL** is a software tool which allows users to quickly and easily create a Relational Database Management System (RDBMS).

- The **punch card** was invented by Joseph Marie Jacquard, and was originally used to automate the textile weaving process. It was later utilized for computer data storage.

- In order to interact with RDBMS, Don Chamberlin and Raymond Boyd developed **Structure Query Language (SQL)**.

“

In the center of bringing any dream into fruition is self-discipline

WILL SMITH

DAY 1

In this chapter, you will:

- Get started with MySQL command line
- Create and drop databases
- Create and drop tables
- Create and delete columns

DAY 1 OVERVIEW

In today's lesson, you will open the MySQL command line and start MySQL. By opening the MySQL command line interface, you are accessing the MySQL RDBMS installed on your computer. Next, you will perform the following tasks:

1) View the contents in your RDBMS
2) Create an empty database
3) Structure an empty table and store it in the database
4) View a description of the table
5) Add new columns to the database
6) View the contents of the table
7) Delete columns from the table
8) Drop the table from the database
9) Drop the database from the RDBMS

This may seem like a lot for Day 1, but most of the commands involve a single line of code.

DATA GURU TIP

SAVE YOUR CODE! The great thing about coding is that you can copy-and-paste code or run a script to execute commands. When you save your code, you don't have to write the same code twice.

SYNTAX RULES

Coding, also called computer programming, allows developers to send instructions to a computer to complete a specific set of actions. When it comes to coding, computers are extremely particular. Each coding or programming language has its own set of rules or instructions called **syntax rules** which must be followed precisely or the code will not run. This section covers general syntax rules in MySQL.

Comments are used to create notes in code without interfering with syntax rules. You can use **/**/** to make long comments in MySQL and **#** or **--** to comment on short lines of code. Comments are great for documenting who the code belongs to, the purpose of the code, and what changes were made to the code over time. Comments make it easier to share code with other developers who may want to use the it or serve as a reminder to the original developer when they haven't viewed the code after a long period of time. When a developer sends a MySQL command to the computer, the comments will be ignored as long as they are denoted in one of the forms depicted on the following page.

Every coding language has a unique set of keywords. **Keywords** cue the computer to expect certain actions or a series of commands. These keywords are often required to be stated in a specific order so that the computer understands and successfully implement the instructions. For example, tables are stored in databases. Developers cannot use keywords to create a table before developing the database to store it in. This action would result in a coding error.

EXAMPLE CODE WITH COMMENTS

```
/*
Code Owner: Jade Wade
Date: 1/27/2020
Purpose: Show the reader how comments
work
*/

/* Displays which databases exist in your
MySQL RDBMS */
SHOW DATABASES;

-- Specify which database to explore
USE DATABASE DATABASE;

# Displays all tables in a database
SHOW TABLES;
```

SYNTAX RULES

There are certain syntax rules in MySQL that are required whenever the programmer has finished giving instructions and is prepared to run their code. After each command statement the developer must type a semicolon and press the Enter key. This alerts the computer to read and process the block of code. If the code is submitted without syntax errors, the message "Query OK" will display on the screen along with other information requested by the user.

If the command does not follow the syntax rules, the user will receive an "ERROR" message. This means that a **bug** or error exists in the code. Don't stress over bugs in your code; this is common in coding and occurs routinely. The process of finding and removing coding errors is called **debugging**. Some developers are hired to test and locate bugs in code. The job title for this function is typically referred to as a *tester*.

DATA GURU TIP

Bugs in code are perfectly normal and they happen often. It is rare to run a block of code the first time without receiving an error. Remember to be patient with yourself and take a break when needed.

SYNTAX CONVENTIONS

The following illustration represents how MySQL code syntax rules are displayed throughout this book. This is the general form of a SQL query, which is covered in Day 3 .

```
SELECT
    Column-1
    <,...Column-n>
FROM DATABASE.TABLE

<WHERE condition>

<GROUP BY Column-1, <...Column-n>>

<ORDER BY Column-1, <...Column-n>>;
```

SELECT, **FROM**, **WHERE**, **GROUP BY**, and **ORDER BY** are displayed in bold white uppercase letters; they must be spelled exactly as displayed.

Column-1, DATABASE, TABLE, and condition are color coded and each represents a value which is provided by the user.

<,...Column-n> is enclosed in angle brackets; it represents optional syntax.

The MySQL syntax presented in this publication is limited to statements needed to complete the exercises throughout this book. For additional syntax, please visit mysql.dn2g.com.

Surface

LESSON 1

GETTING STARTED

MYSQL COMMAND LINE

Most servers have a **Command Line Interface (CLI)** or terminal in which the user can submit instructions to the computer. If you work in the IT industry, there is a high probability you will not always have the luxury of using a pretty, point-and-click operating system (OS) interface such as Windows or macOS. In this book, you will prepare for this scenario which occurs often in the tech industry, by using MySQL via a CLI. Working with a CLI prepares you to connect and work with various company servers with ease.

LESSON 1: GENERAL COMMANDS

```
/* Displays which databases exist in your
MySQL RDBMS */
SHOW DATABASES;

/* Create a new database which does not
already exist in your MySQL RDBMS */
CREATE DATABASE DATABASE;

-- Specify which database to explore
USE DATABASE DATABASE;

/* Display all tables which exist in a
database */
SHOW TABLES;

/* Provides a description of a specified table
including column names and data types */
DESCRIBE DATABASE.TABLE;
```

LESSON 1.1

1) Open the MySQL 8.0 Command Line Client.

2) A black screen will appear with the prompt "Enter password:". Type in the Root Account Password you set during installation and press the Enter key.

The system will display: the MySQL welcome message, copyright information, and a few command shortcuts (i.e., help and clear). You should see your cursor blinking next to **mysql>**. This indicates that MySQL is waiting for you to enter a command.

3) Type the command **SHOW DATABASES;** and press the Enter key. You will see a list of databases currently installed on your MySQL server as shown on the next page. **Note: Do not delete or edit these databases which are standard in MySQL software as this action may negatively impact the integrity of the database system.**

SYNTAX TIP

MySQL is case insensitive when it pertains to system commands, keywords, and column names. For example, the following queries would run successfully and result in the same output:

```
show databases;
SHOW DATABASES;
SHOW databases;
shoW daTABases;
```

In this book, all commands follow the same case formatting and color code style to assist with learning and readbility.

ANSWER 1.1

MySQL 8.0 Command Line Client

```
Enter password: **********
Welcome to the MySQL monitor.  Commands end with ; or \g.
Your MySQL connection id is 10
Server version: 8.0.17 MySQL Community Server - GPL

Copyright (c) 2000, 2019, Oracle and/or its affiliates. All rights reserve

Oracle is a registered trademark of Oracle Corporation and/or its
affiliates. Other names may be trademarks of their respective
owners.

Type 'help;' or '\h' for help. Type '\c' to clear the current input statem

mysql> SHOW DATABASES;
+--------------------+
| Database           |
+--------------------+
| information_schema |
| mysql              |
| performance_schema |
| sys                |
+--------------------+
4 rows in set (0.00 sec)

mysql>
```

LESSON 1.2

To being storing data in an RDBMS, the developer must first create a place to store it. In this lesson, you will create a new database called TEST_DATABASE. Use the Lesson 1 General Commands and follow the steps below to create a the database.

1) In the command line type **CREATE DATABASE TEST_DATABASE;** and then press the Enter key.

2) To see if your database was successfully created, type **SHOW DATABASES;** and press the Enter key. You should see your database listed on the screen.

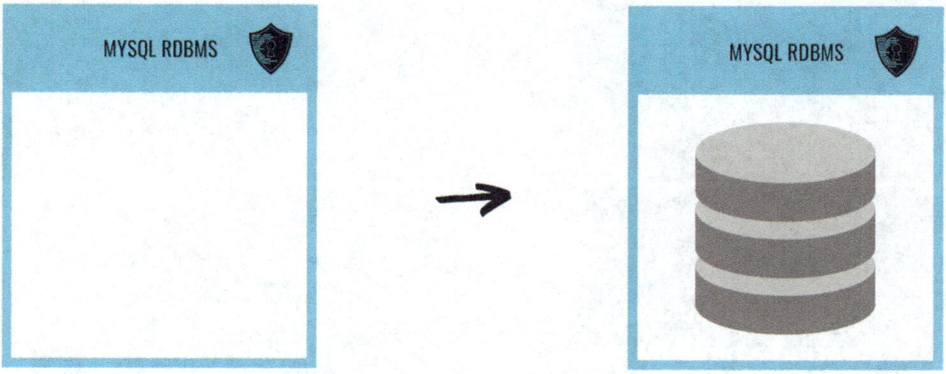

ANSWER 1.2

MySQL 8.0 Command Line Client

```
mysql> CREATE DATABASE TEST_DATABASE;
Query OK, 1 row affected (0.09 sec)

mysql> SHOW DATABASES;
+--------------------+
| Database           |
+--------------------+
| information_schema |
| mysql              |
| performance_schema |
| sys                |
| test_database      |
+--------------------+
5 rows in set (0.00 sec)

mysql>
```

DATA GURU TIP

You cannot use the name of an existing database when creating a new database. If you do, you will receive an error when attempting to overwrite the existing database. If you desire to overwrite an existing database, you must delete it first and create a new one.

TABLE CREATION OVERVIEW

In order to create an empty table, the user must define a schema. A **schema** provides ordered structure to database tables and specifies rules for the data types assigned in each column. In a relational table, users are required to setup a schema before adding data to the table. To accomplish this, the developer must write schema code to inform the database of the following:

- The name of the table;
- The name of each field or column in the table;
- The data types that can be stored in each column;
- Which columns are primary or secondary keys.

Next, the table schema is placed in a database. In this exercise it will be placed in database **TEST_DATABASE**. You'll create a schema for a table named **TEST** that has one column. Don't worry if you don't understand all the data types and keywords yet. They will be covered in depth on Day 2. The important thing is for you to become familiar with and understand the sequential steps required to create a table.

TABLE SCHEMA CREATION SYNTAX

```
-- Create a new table schema
CREATE TABLE TABLE
    (
        Column-1 datatype-1
        <,... Column-n datatype-n>

    );
```

LESSON 1.3

In this lesson, you will walk through the steps of creating a table called TEST in your new database TEST_DATABASE.

1) In the MySQL command line, type **USE TEST_DATABASE;** and press the Enter key. This signals the computer to start using your new database until you specify that you want to use another database or close the program.

2) Type **CREATE TABLE TEST** and press Enter.

3) On a new line type **(COL1 INT);** and press Enter. You can type commands on separate lines as shown in the example on the next page or leave them on one line. Either option works well in MySQL.

If your code has been entered correctly, at the bottom of your screen you should see the message "Query OK. 0 rows affected" followed by the length of time required for your computer to execute the command.

You should have a table named TEST in your TEST_DATABASE database. You can check by typing in the command **SHOW TABLES;** followed by the Enter key. TEST should now be in your table list.

ANSWER 1.3

MySQL 8.0 Command Line Client

```
mysql> CREATE TABLE TEST
    -> (
    -> COL1 INT
    -> );
Query OK, 0 rows affected (0.44 sec)

mysql> SHOW TABLES;
+--------------------------+
| Tables_in_test_database |
+--------------------------+
| test                     |
+--------------------------+
1 row in set (0.00 sec)
```

DATA GURU TIP

When you have long commands, separate them by entering your code on different lines using the Enter key. This makes your code organized, readable, and easier to debug.

ALTER TABLES OVERVIEW

After creating a table, you might have to revise it by adding or removing one or more columns to the table. One way to make changes to an existing dataset, or table, is to use the **ALTER TABLE** keywords. In the last section, you created an empty table with one column. In the next lesson, you will learn to add a new column to the table using the MySQL syntax on the following page.

To add new columns to an existing table the developer will need he following information:

- The name of the table;
- The name of each field or column in the table;
- The data types that can be stored in each column.

When removing columns, the user only needs the names of the columns they wish to delete.

DATA GURU TIP

When completing projects that require you to create a table, take time to map out and plan your table schemas. Try to think of various scenarios in which you will use your table and ensure you have all of the columns you need.

ALTER & DELETE SYNTAX

```
/* Add one or more new columns to an
existing database*/
ALTER TABLE TABLE
    ADD COLUMN
        Column-1 datatype-1
        <,... Column-n datatype-n>
    AFTER ExistingColumn;

/* Drop one or more columns from an
existing database*/
ALTER TABLE TABLE
    DROP COLUMN
        ExistingColumn;

-- Delete a table
DROP TABLE TABLE;

-- Delete a database
DROP DATABASE DATABASE;
```

LESSON 1.4

Now you will add a new column named COL2 to your TEST table. Type the **DESCRIBE TEST;** command and press Enter to see the current column names and data types which exist inside of your table before submitting the query below.

1) First type **ALTER TABLE TEST** then press the Enter key to notify the computer that you will be modifying the specified table.

2) Type **ADD COLUMN** and press the Enter key to move to the next line.

3) Next, type **COL2 INT** and press the Enter key to indicate that you are adding the new column.

4) Lastly type **AFTER COL1;** and press the Enter key to signal to the computer that the new column will be placed after COL1 in the table.

To see if your command was successful, type the **DESCRIBE TEST;** command and press Enter again. You should see the new column name with the corresponding data type in your table.

ANSWER 1.4

MySQL 8.0 Command Line Client

```
mysql> DESCRIBE TEST;
+-------+---------+------+-----+---------+-------+
| Field | Type    | Null | Key | Default | Extra |
+-------+---------+------+-----+---------+-------+
| COL1  | int(11) | YES  |     | NULL    |       |
+-------+---------+------+-----+---------+-------+
1 row in set (0.00 sec)

mysql> ALTER TABLE TEST
    -> ADD COLUMN COL2 INT AFTER COL1;
Query OK, 0 rows affected (0.24 sec)
Records: 0  Duplicates: 0  Warnings: 0

mysql> DESCRIBE TEST;
+-------+---------+------+-----+---------+-------+
| Field | Type    | Null | Key | Default | Extra |
+-------+---------+------+-----+---------+-------+
| COL1  | int(11) | YES  |     | NULL    |       |
| COL2  | int(11) | YES  |     | NULL    |       |
+-------+---------+------+-----+---------+-------+
2 rows in set (0.00 sec)
```

REMOVING DATA

Imagine you are the owner of a successful online shop. Over a five year period, your business has decided to stop selling a few unpopular products and replace them with new high demand items. One day, you notice that you have a database that is reaching it's storage capacity. You consider purchasing additional servers to store data, but think it may be smarter to save money and remove the data you no longer need. This is one of many instances in which you may elect to remove data, such as the unpopular products, from your RDBMS.

Now that you've learned how to create empty tables and databases. It is time to learn how to remove them. There may be instances when you no longer need the data and deleting information will become a necessity. **However, be careful with these commands and use them wisely. It is extremely hard to recover data once it has been deleted.** You will use the *Alter & Delete Syntax* to practice eliminating data.

LESSON 1.5

In this lesson, you will remove COL2 from your TEST table. Type the **DESCRIBE TEST;** command and press Enter to see the current column names and data types that exist in your table before submitting the query below.

1) First,type **ALTER TABLE TEST,** then press the Enter key to notify the computer that you are modifying the specified table.

2) Type **DROP COLUMN COL2;** and press the Enter key to remove the column from the table.

To see if your command was successful, type the **DESCRIBE TEST;** command and press Enter. If your command was successful, COL2 will no longer be present in the TEST table.

ANSWER 1.5

MySQL 8.0 Command Line Client

```
mysql> DESCRIBE TEST;
+-------+---------+------+-----+---------+-------+
| Field | Type    | Null | Key | Default | Extra |
+-------+---------+------+-----+---------+-------+
| COL1  | int(11) | YES  |     | NULL    |       |
| COL2  | int(11) | YES  |     | NULL    |       |
+-------+---------+------+-----+---------+-------+
2 rows in set (0.00 sec)

mysql> ALTER TABLE TEST
    -> DROP COLUMN COL2;
Query OK, 0 rows affected (0.54 sec)
Records: 0  Duplicates: 0  Warnings: 0

mysql> DESCRIBE TEST;
+-------+---------+------+-----+---------+-------+
| Field | Type    | Null | Key | Default | Extra |
+-------+---------+------+-----+---------+-------+
| COL1  | int(11) | YES  |     | NULL    |       |
+-------+---------+------+-----+---------+-------+
1 row in set (0.00 sec)
```

LESSON 1.6

In this exercise, you will remove the **TEST** table from the **TEST_DATABASE**. Type the **SHOW TABLES;** command and press Enter to view all the tables in the database.

Type **DROP TABLE TEST;** and press Enter to remove the empty table you previously created. Then, type the **SHOW TABLES;** command and press Enter. You should no longer see the table in the database. You will see a message on the screen stating "Empty set" because the database is empty.

See how easy it is to drop a table from a database? Before you remove information be sure that this is the action you want to take.

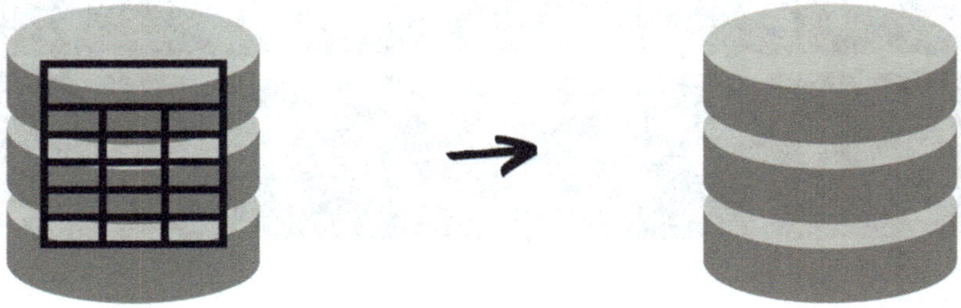

ANSWER 1.6

MySQL 8.0 Command Line Client

```
mysql> SHOW TABLES;
+---------------------------+
| Tables_in_test_database |
+---------------------------+
| test                      |
+---------------------------+
1 row in set (0.12 sec)

mysql> DROP TABLE TEST;
Query OK, 0 rows affected (0.36 sec)

mysql> SHOW TABLES;
Empty set (0.00 sec)
```

LESSON 1.7

Finally, you will remove your TEST_DATABASE from the MySQL RDBMS. Type the **SHOW DATABASES;** command and press Enter to see which databases currently exist in your RDBMS.

Type **DROP DATABASE TEST_DATABASE;** and press Enter to remove it. Then, type the **SHOW DATABASE;** command again to see that the database is no longer present.

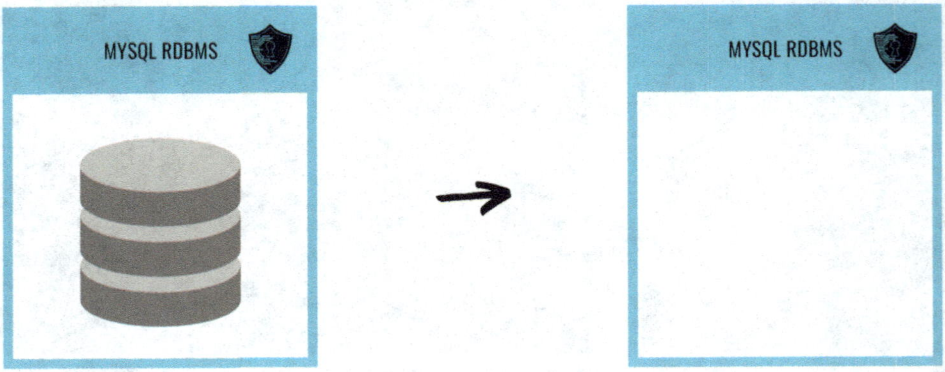

ANSWER 1.7

MySQL 8.0 Command Line Client

```
mysql> SHOW DATABASES;
+--------------------+
| Database           |
+--------------------+
| information_schema |
| mysql              |
| performance_schema |
| sys                |
| test_database      |
+--------------------+
5 rows in set (0.00 sec)

mysql> DROP DATABASE TEST_DATABASE;
Query OK, 0 rows affected (0.07 sec)

mysql> SHOW DATABASES;
+--------------------+
| Database           |
+--------------------+
| information_schema |
| mysql              |
| performance_schema |
| sys                |
+--------------------+
```

SUMMARY

In this chapter you learned:

- You can use **/**/** to make long comments in MySQL and **#** or **--** to comment short lines of code

- **Syntax rules** are used to provide exact instructions to the computer and **bugs** occur when these syntax rules aren't followed

- **SHOW DATABASES** keywords are used to view the databases which currently exist in an RDBMS

- **USE DATABASE** keywords specify which database to explore and manage

- **CREATE DATABASE** command is used to create a database and the **CREATE TABLE** command is used to create a table or table schema

- **DESCRIBE TABLE** allows the user to view the column names and corresponding data types in a table

- **ALTER TABLE** command is used to make changes to an existing table

- **DROP DATABASE** command is used to remove a database and the **DROP TABLE** command is used to remove a table

“

If you double the number of experiments you do per year,

You’re going to double your inventiveness

JEFF BEZOS
CEO OF AMAZON

DAY 2

In this chapter, you will:

- Learn how data is imported into databases
- Get familiar with common data types in computer programming
- Enter data into tables manually
- Import data into tables from a CSV file
- Setup a database to prepare for Day 3

DAY 2 OVERVIEW

There are three major procedures that developers should understand when managing data:

1) Importing data into a database;
2) Manipulating data; and
3) Exporting data from a database

The content in Day One covered the creation of databases and table schemas. This chapter details how to import data into a table. There are several techniques that can used to import data into a table, but most begin with a form of manual entry. The manual entry process is critical for businesses which is evident by the vast number of available jobs and established companies that revolve around the data entry process alone.

If you've filled out a digital registration form or a survey, you've likely typed data into the form manually. The data was stored in a table or dataset once the form was completed and submitted. The lessons in this chapter reveal just how data is stored after it is entered into a digital form.

Truthfully, manual data entry is quite tedious, but the data entry lessons are kept short to prevent fatigued fingers. This chapter details how to import **comma separated value (CSV)** files into a database table which give developers the option of entering data in a more automated fashion. **CSV** files are a type of text document in which all the data elements are separated or delimited by commas. Often, the files have header columns on the first row labeled with column names. The rest of the file contains row entries and each element is separated by a comma.

CSV FILE EXAMPLE

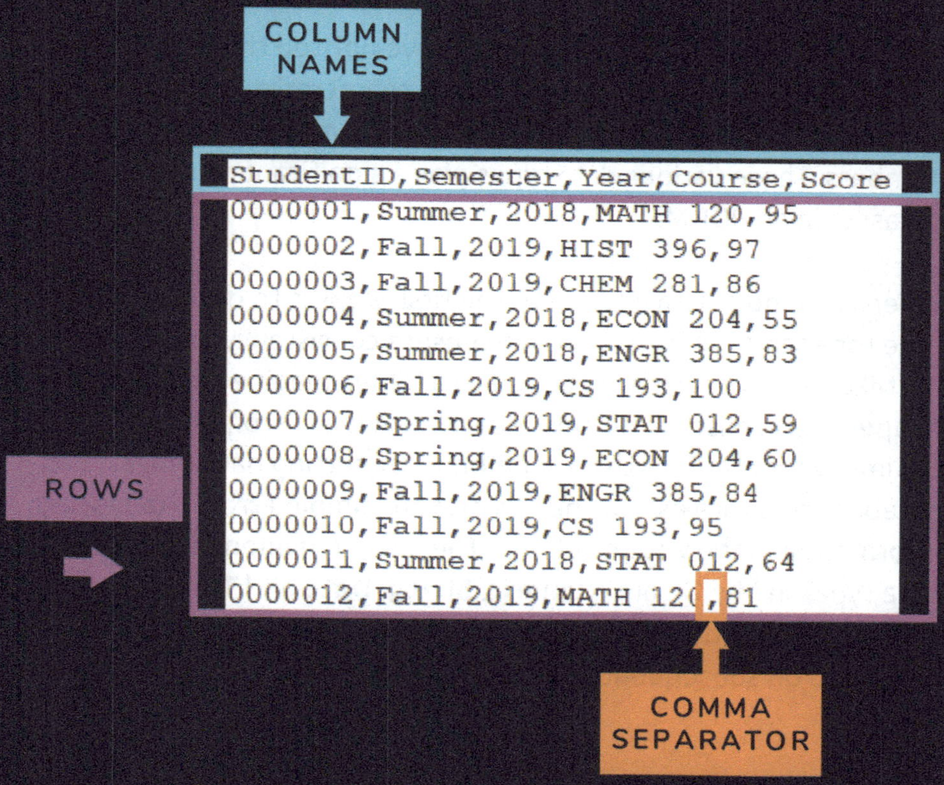

DATA GURU TIP

As humans it is easy to make a mistake, even when composing a short email. Manual data entry is no different, you will find issues with most raw datasets. Don't be alarmed; this is normal. A common term used for data discrepencies is **dirty data**.

DATA TYPES

In previous table examples, you may have noticed that the data contained in each column has different characteristics . For instance, some columns have dates, while other are comprised of words or numbers. These various data types play a significant role in creating datasets in an RDBMS.

When creating a dataset, one of the first steps is to define the data type for each field. This is done to ensure consistency and prevent mistakes when new data is added to a table. Imagine what would happen if you mistakenly entered and saved a customer's name in a column where their birth date belongs. By setting data types, database rules are established to prevent simple errors like placing information in the wrong column. The most common categories of data types in MySQL are **Numeric**, **String**, **Date** and **Time**. There are many variations of the aforementioned data types which will be described on subsequent pages.

DATA GURU TIP

Have you ever attempted to store files on a computer or flash drive and received the message that you are low on space? This happens with databases too. Each data type reserves a set amount of space for data storage. **Bits** are units of digital information used to describe the amount of storage on a computer. **Bytes** are a standard digital storage metric which represents 8 bits. If you need to conserve on space utlization in your MySQL server, refer to the MySQL developer documents to determine the number of bytes each data type uses.

NUMERIC DATA TYPES

Numeric data types are used to store quantifiable values represented by numbers. Here are a few typical examples of values often classified as numeric data types in programming:

Distance in Miles/Kilometers
Height in Inches/Centimeters
Product Prices
Household Size

These examples represent quantities or measurable objects in which a mathematical equation is appropriate. For instance, obtaining the average household size or adding multiple product prices are reasonable calculations that provide useful information.

Values also exist that appear to be numeric, but they are actually labels or descriptive values and classified as string data types. For example, the following data types use numbers to represent them, but they are not measurable:

Zip Codes
Passport Identification Number
Bar Code Number

Adding zip codes or dividing Passport ID numbers, doesn't make much sense. That's because these numbers symbolize persons, places, or things and should not be used as numeric data types when defining columns that contain values such as these.

INTEGERS

In computer programming, numeric data types are often represented as either whole or decimal numbers. The term **integers** refers to whole numbers that do not contain decimal points, represent fractional values, or other imaginary numbers. Integers can be positive, zero, or negative numbers.

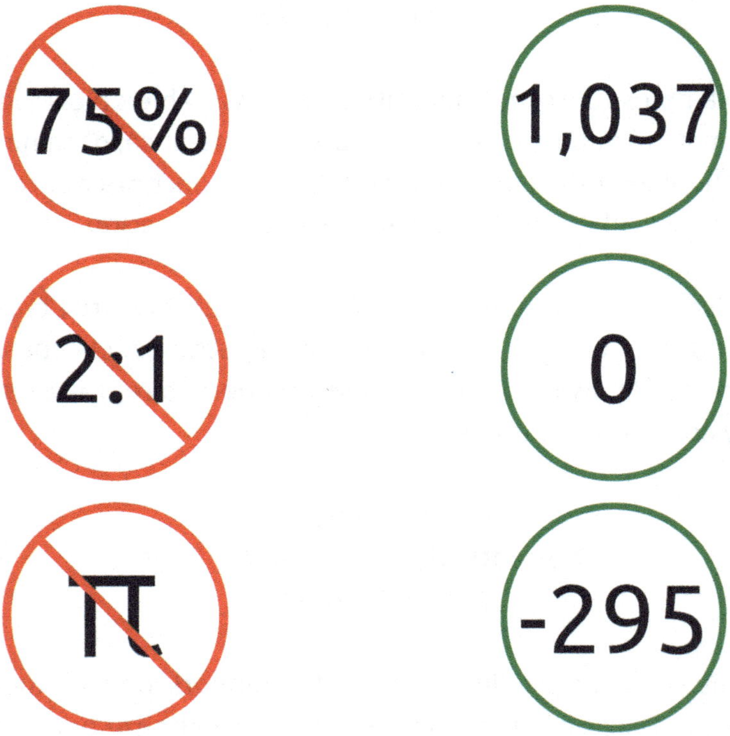

MySQL currently has five commands that are used to define columns that contain integers: **TINYINT**, **SMALLINT**, **MEDIUMINT**, **INT**, and **BIGINT**. You may be pondering, *Why are there so many different variations of integers in MySQL?* Each integer variant reserves a predetermined amount of storage and has unique limitations on the minimum and maximum value that can be stored in a corresponding column.

INTEGER TYPE CONSTRAINTS

TYPE	STORAGE (BYTES)	MINIMUM VALUE	MAXIMUM UNSIGNED
TINYINT	1	-128	255
SMALLINT	2	-32768	65535
MEDIUMINT	3	-8388608	16777215
INT	4	-2147483648	4294967295
BIGINT	8	-2^{63}	2^{64}

All data types have limitations in programming. For instance, at the time this book was written *TINYINT* data types in MySQL only allows values between -128 and 255. This means developers will not be able to store a number lower than -128, such as -129, in a column set to accept *TINYINT* values. To find the most recent storage and limitations for all data types, please refer to the MySQL developer documents at dtypes.dn2g.com

DECIMALS

Decimal points allow us to view numbers with more precision. Precision represents the amount of information communicated about a number. The greater the quantity of digits that exist following the decimal point, the more precise the number.

2.999999827 is more precise than 2.99
2.99 is more precise than 3

The **DECIMAL(*precision, scale*)** command is used to define data types which contain decimals. This command accepts two inputs:

- ***Precision*** which limits how many digits the overall number can include and
- ***Scale*** which restricts the number of digits placed behind the decimal point.

A column with a defined data type as *DECIMAL(4,2)* can only contain a total of 4 digits including the two digits following the decimal point (i.e. 35.57 or 18.94). If you attempt to store *42.64236* in this column, the number is rounded at the 2nd digit behind the decimal point and the remaining three digits are dropped. This number is stored as *42.64*, discarding the remaining three digits of precision.

STRINGS - CHARS & VARCHARS

A **character** is a letter, digit, or special character that is used to form a word or used as a separator. A sequence of characters is referred to as a **string**. In computer programming strings and characters are signified with single quotes or double quotes placed at the beginning and end of the string. If you choose to denote a string with single quotes, it must begin and end with a single quote (the same rationale applies to double quotes). Below are examples of how characters and strings are formatted when stored in columns.

'y'
'yes hello'
"hello world!"
"@username"

MySQL has eight different string data type commands, but only two prevailing data types are covered in this chapter: **CHAR(*length*)** and **VARCHAR(*length*)**. *CHAR* and *VARCHAR* both represent strings, but they store data differently and have varying limitations. When typing either of the two commands, you must specify the desired maximum length of your string using the ***length*** input parameter.

DATA GURU TIP

If one of your string data elements contains a comma? Then a CSV file may not be the best type of file to store and import your data. There are other types of separators or delimiters including tabs, pipes, semicolons, etc. For example, files which contain tab separated values are referred to as **TSV**. For more examples on TSV files checkout tsv.dn2g.com

STRINGS - CHARS & VARCHARS

A column with a data type defined as *CHAR(5)* or *VARCHAR(5)* means the string stored in the column can only be up to five characters in length. However, if the string input into the column is less then five characters, the *CHAR* data types will pad the empty area with **whitespace** or blanks. On the other hand, *VARCHAR* data types do not pad columns with whitespace. See the table below which compares the storage of strings when they are less than the maximum specified length. Note: Attempting to store strings greater than the set limit, will result in removal or **truncation** of the remaining characters.

VALUE	CHAR (4)	STORAGE REQUIRED	VARCHAR(4)	STORAGE REQUIRED
''	' '	4 bytes	''	1 byte
'ab'	'ab '	4 bytes	'ab'	3 bytes
'abcd'	'abcd'	4 bytes	'abcd'	5 bytes
'abcdefgh'	'abcd'	4 bytes	'abcd'	5 bytes

The length of a *CHAR* data value can be between 0 to 255 characters. However, if more characters are needed, use the *VARCHAR* data type instead. VARCHAR string lengths are restricted to the remaining bytes of storage in the row. Note: Each row in a MySQL table is limited to 65,535 bytes of storage.

Earlier in this chapter, you read about the existence of values which appear to be numeric, but are used as labels (i.e. Zip Codes, ID Numbers, etc). These are instances in which you would use the *CHAR* and *VARCHAR* commands to specify the column data type.

DATES & TIMES

The last category of data types covered in this chapter are Dates, Times, and Timestamps. The values stored in these columns typically represent the time of an event occurrence. If you operated a large store, your company database might contain many date and time related columns in each table, such as:

Timestamp of Product Ordered
Date New Employee was Hired
Year Customer Signed Up for Membership
Time Employee Clocked in for Work

There are five MySQL commands used to define date and time data types: **DATE**, **TIME**, **DATETIME**, **TIMESTAMP**, and **YEAR**. The default formats for date and time data types are shown below.

DATA TYPE	FORMAT
DATE	'2020-01-01'
TIME	'08:30:55'
DATETIME	'2020-01-01 08:30:55'
TIMESTAMP	'2020-01-01 08:30:55'
YEAR	2020

This book introduces only the most commonly used data types in computer programming and demonstrates how they are used in MySQL. In the lessons at the end of this chapter, only the data types covered are needed to enter data elements into a database. To learn more about other MySQL data types, consult the MySQL developer documents at dtypes.dn2g.com.

IMPORTING DATA

THE FIRST STEP

Setting up databases and importing data can be a tedious process. Typically, in large organizations there are multiple IT teams dedicated to ensuring this process runs smoothly. This process is referred to as **Extract, Transform, and Load (ETL)**. ETL developers are responsible for making sure:

- database servers can support growing data storage needs;
- data is imported in a timely manner;
- quality data is being stored in the database; and;
- table connections and data descriptions are documented

These are just a few responsibilities to keep in mind when designing database systems.

LESSON 2.1

In today's lesson, you will create a new table schema using the specifications from the table below. Next, you will enter data manually and import data into a table from a file.

First, you must create a database to store your new table. This activity was covered in Day 1. Recreate the database TEST_DATABASE by repeating the steps below from Lesson 1.2.

1) Type the command **CREATE DATABASE TEST_DATABASE;** and press the Enter key.

2) Next, type **SHOW DATABASES;** and press the Enter key. You should see the database you created, TEST_DATABASE, in the list of databases on your screen as shown in the example on the next page.

COLUMN NAME	DATA TYPE	MYSQL KEYWORD
RowID	String	VARCHAR(6)
Name	String	VARCHAR(50)
BirthDate	Date	DATE
FavoriteFood	String	VARCHAR(20)
LuckyNumber	Number	TINYINT

ANSWER 2.1

MySQL 8.0 Command Line Client

```
mysql> CREATE DATABASE TEST_DATABASE;
Query OK, 1 row affected (0.09 sec)

mysql> SHOW DATABASES;
+--------------------+
| Database           |
+--------------------+
| information_schema |
| mysql              |
| performance_schema |
| sys                |
| test_database      |
+--------------------+
5 rows in set (0.00 sec)

mysql>
```

LESSON 2.2

Now you will create a new table schema called TEST. You will use the same table schema shown on page 81.

1) Type **USE TEST_DATABASE;** and then press the Enter key.

2) To define the table schema, type the command **CREATE TABLE TEST** and press the Enter key.

3) Next, you will type an open parenthesis **(**.

4) Now type the column name, a space, and the corresponding data type as shown in the example on the next page. Type a comma (**,**) after the data type and press the Enter key.

5) Repeat Step 4 until all the information for the table schema has been entered. Once you get to the last data type, do not add a comma to the end. Instead, type a closed parenthesis followed by a semicolon **);** then, press the Enter key.

Now, you should have a table named TEST in your TEST_DATABASE database. You can check by typing the command **SHOW TABLES;** followed by the Enter key. TEST should be in your list of tables.

In Lesson 2.3, you will add rows of data to the new table schema.

ANSWER 2.2

MySQL 8.0 Command Line Client

```
mysql> CREATE TABLE TEST
    -> (
    -> RowID VARCHAR(6),
    -> Name VARCHAR(50),
    -> BirthDate DATE,
    -> FavoriteFood VARCHAR(20),
    -> LuckyNumber TINYINT
    -> );
Query OK, 0 rows affected (0.59 sec)

mysql> SHOW TABLES;
+-------------------------+
| Tables_in_test_database |
+-------------------------+
| test                    |
+-------------------------+
1 row in set (0.00 sec)

mysql>
```

DATA GURU TIP

When you have long commands, separate them on different lines using the Enter key. This makes your code more readable, it is easier to debug, and reusable in the future.

INSERTING ROWS OVERVIEW

At this point, the **TEST** table should be empty with only the columns labeled. You're now ready to start adding data to the table.

In relational database tables, one way to insert data into a table is by using the **INSERT INTO** statement. The syntax to accomplish this is shown on the next page. The user first tells the computer which table to put the data in. The next step is to specify the column order the values will be placed into the table. Lastly, values are entered row by row which is indicated by an open and closed parentheses. The values in each row are separated by a comma.

The *INSERT INTO* command allows you to add rows to both empty tables or tables which already exist. You must ensure that the data you are adding corresponds to the order of your columns and its data type. If you were to mistakenly enter the string"Apple Pie' under the **LuckyNumber** column, you would receive an error and the command would not execute. In the next lessons, you will learn how to add rows of data to your **TEST** table using the *INSERT INTO* command syntax shown on the following page.

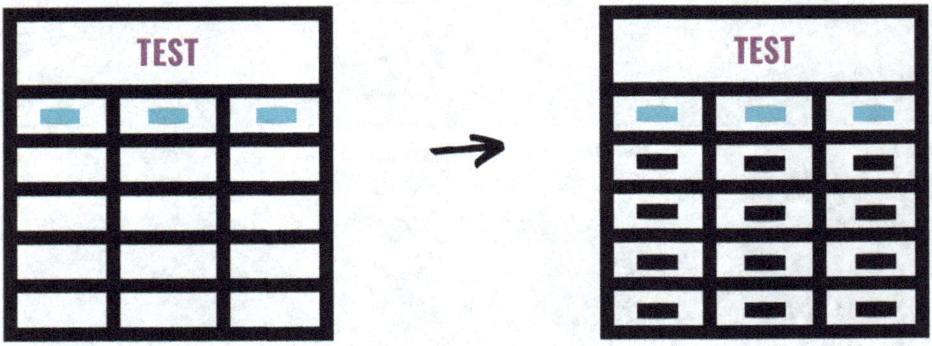

INSERT ROWS SYNTAX

```
--Syntax for inserting data into a table
INSERT INTO TABLE
  (
     Column-1
     <,...Column-n>
  )

  VALUES
     value-1
     <,...value-n>;
```

LESSON 2.3

Let's add data to the **TEST** table using the *INSERT INTO* statement. Choose a few names, dates, food items, and numbers to insert into the table. Also, select a unique row ID in the form of a 6-digit character string to use for tracking purposes.

1) Type the statement **INSERT INTO TEST (** and press Enter.

2) Using the specifications in the table on page 81, type the column names followed by a comma (,) and press the Enter key.

3) Repeat Step 2 until all the column names have been entered. After the last column *LuckyNumber TINTYINT* has been entered, do not add a comma at the end. Press the Enter key. Next, type a closed parenthesis and press the Enter key.

4) Next type the **VALUES** command which alerts the computer to expect information to be inserted therow. Then press the Enter key.

5) Next, enter the an open parentheses **(** followed by a unique Row ID, Name, Date, Food Item, and a Number. Then add a closed parentheses **)**, followed by a comma (**,**). Note: Make sure all values have the expected data type formats shown in table on page 81.

6) Repeat Step 5 until all desired row values have been entered. Once you get to the last row, do not add a comma at the end. Instead, add a semicolon (**;**) and press the Enter key.

To see if the rows were added to the table, use a *SELECT* statement. Simply type **SELECT * FROM TEST;** to view all of the data currently stored in the **TEST** table. The *SELECT* statement will be covered in detail in tomorrow's lesson.

ANSWER 2.3

MySQL 8.0 Command Line Client

```
mysql> INSERT INTO TEST
    -> (
    -> RowID,
    -> Name,
    -> BirthDate,
    -> FavoriteFood,
    -> LuckyNumber)
    -> VALUES
    -> ("000001", "Jade", "1995-09-24", "Apple Pie", 27),
    -> ("000002", "Elliot", "2001-02-22", "Dried Squid", 95);
Query OK, 2 rows affected (0.11 sec)
Records: 2  Duplicates: 0  Warnings: 0

mysql> SELECT * FROM TEST;
+--------+--------+------------+--------------+-------------+
| RowID  | Name   | BirthDate  | FavoriteFood | LuckyNumber |
+--------+--------+------------+--------------+-------------+
| 000001 | Jade   | 1995-09-24 | Apple Pie    |          27 |
| 000002 | Elliot | 2001-02-22 | Dried Squid  |          95 |
+--------+--------+------------+--------------+-------------+
2 rows in set (0.00 sec)
```

IMPORTING CSV FILES

Go to testdata.dn2g.com and click the ⭳ icon to download the CSV file labeled ***add_test_data.csv*** which contains the information shown below. The first row in this file contains the column names shown in the **TEST** table. In the next lesson, you will add this data to your **TEST** table using the syntax on the following page.

```
RowID,Name,BirthDate,FavoriteFood,LuckyNumber
000003,Erin,1981-11-09,Tuna Pasta,22
000004,Alyce,1984-07-20,Adobo,35
000005,Nobuko,1984-09-11,Hamburger,19
000006,Yee,1985-07-20,Pizza,23
000007,Chante,1985-11-09,Barbecue,60
000008,Vivian,1986-10-09,Tacos,28
000009,Branda,1987-02-11,Banana split,11
000010,Gisel,1987-05-13,Fried Chicken,5
```

To import the file using MySQL, you will need the CSV file location on your computer where it was downloaded and stored. The directory and file name should look similar to the format below:

'C:/temp/add_test_data.csv'

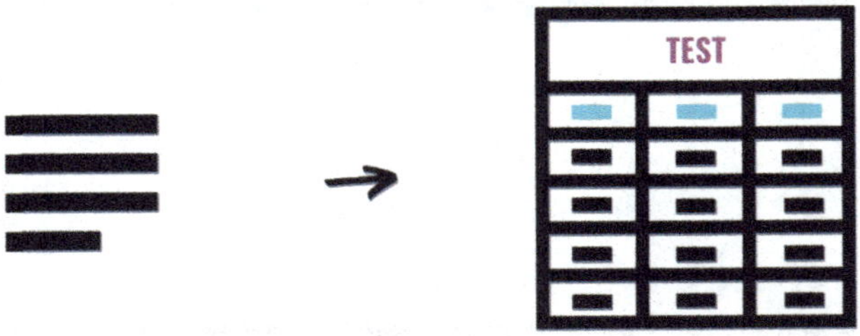

IMPORT FILE SYNTAX

```
-- Load data into a table from a file
LOAD DATA INFILE '/directory/file'

INTO TABLE TABLE

FIELDS TERMINATED BY 'delimiter'

ENCLOSED BY string_identifier

LINES TERMINATED BY
      newline_identifier

<IGNORE 1 ROWS>;
```

LESSON 2.4

1) Now that you have the file location, start by typing the **LOAD DATA INFILE** command followed by the file location enclosed in single quotes. Then press the Enter key.

2) Next, type **INTO TABLE TEST** so the computer will know which table to store the data in. Then press the Enter key.

3) Next, indicate that the values in the file are separated by commas by typing **FIELDS TERMINATED BY ','** then press the Enter key.

4) Next, specify that all string values in the file are enclosed in double quotes by typing **ENCLOSED BY '"'** and then press Enter.

5) The computer needs to know how to determine one row from the next. Each row of data in the CSV file is on a separate line, so type **LINES TERMINATED BY '\n'** to indicate the newline character is where the row ends and then press Enter.

6) The CSV file contains column names in the first row, but the column names are already present in the table, therefore, we will skip the first row so it isn't imported. Type **IGNORE 1 ROWS;** and press the Enter key.

To see if the rows were added to the table, type **SELECT * FROM TEST;** and view all of the data currently stored in the **TEST** table.

ANSWER 2.4

MySQL 8.0 Command Line Client

```
mysql> LOAD DATA INFILE 'C:/temp/sqldataguru/add_test_data.csv'
    -> INTO TABLE TEST
    -> FIELDS TERMINATED BY ','
    -> ENCLOSED BY '"'
    -> IGNORE 1 ROWS;
Query OK, 8 rows affected (0.14 sec)
Records: 8  Deleted: 0  Skipped: 0  Warnings: 0

mysql> SELECT * FROM TEST;
+--------+--------+------------+---------------+-------------+
| RowID  | Name   | BirthDate  | FavoriteFood  | LuckyNumber |
+--------+--------+------------+---------------+-------------+
| 000001 | Jade   | 1995-09-24 | Apple Pie     |          27 |
| 000002 | Elliot | 2001-02-22 | Dried Squid   |          95 |
| 000003 | Erin   | 1981-11-09 | Tuna Pasta    |          22 |
| 000004 | Alyce  | 1984-07-20 | Adobo         |          35 |
| 000005 | Nobuko | 1984-09-11 | Hamburger     |          19 |
| 000006 | Yee    | 1985-07-20 | Pizza         |          23 |
| 000007 | Chante | 1985-11-09 | Barbecue      |          60 |
| 000008 | Vivian | 1986-10-09 | Tacos         |          28 |
| 000009 | Branda | 1987-02-11 | Banana split  |          11 |
| 000010 | Gisel  | 1987-05-13 | Fried Chicken |           5 |
+--------+--------+------------+---------------+-------------+
10 rows in set (0.00 sec)
```

DAY 3 PREP

Tomorrow, you will learn how to create reports using a new database. To install this database follow the steps below:

1) Go to the Gitlab repository to obtain the need files by clicking or typing the following URL into you web browser dataguru.dn2g.com and click the ⬇ icon to download the ***dataguru.sql*** file.

2) Once the file finishes downloading, note the folder location of the new file. In the MySQL 8.0 Command Client, type the command **SOURCE**, followed by a space and the file location. Then type **dataguru.sql;** and press the Enter key. Your command should look something like this:

SOURCE C:/temp/dataguru.sql;

The *SOURCE* keyword prompts the computer to run a series of MySQL commands contained in the subsequent file. If you view the *dataguru.sql* file on Gitlab, you will see the commands inside the *dataguru.sql* file and explore it.

After the above command has been run successfully, check to see if a new **DATAGURU** database was successfully created by typing the **SHOW DATABASES;** command and pressing enter. Next type **USE DATATGURU;** followed by the Enter key. Then type, **SHOW TABLES;** and press the Enter key. You should see the **CAREER**, **GRADES**, **FACULTY**, **NEWFACULTY**, and **STUDENTS** tables in your database.

After successful completion of this step, you are prepared for tomorrow's lesson!

MySQL 8.0 Command Line Client

```
mysql> source C:/temp/dataguru.sql
Query OK, 1 row affected (0.13 sec)

Database changed
Query OK, 0 rows affected (0.52 sec)

Query OK, 100 rows affected (0.06 sec)
Records: 100  Duplicates: 0  Warnings: 0

Query OK, 0 rows affected (0.55 sec)

Query OK, 101 rows affected (0.07 sec)
Records: 101  Duplicates: 0  Warnings: 0

Query OK, 0 rows affected (0.43 sec)

Query OK, 98 rows affected (0.08 sec)
Records: 98  Duplicates: 0  Warnings: 0

Query OK, 0 rows affected (0.51 sec)

Query OK, 20 rows affected (0.10 sec)
Records: 20  Duplicates: 0  Warnings: 0

Query OK, 0 rows affected (0.57 sec)

Query OK, 23 rows affected (0.18 sec)
Records: 23  Duplicates: 0  Warnings: 0
```

MySQL 8.0 Command Line Client

```
mysql> SHOW DATABASES;
+--------------------+
| Database           |
+--------------------+
| dataguru           |
| information_schema |
| mysql              |
| performance_schema |
| sys                |
| test_database      |
+--------------------+
6 rows in set (0.00 sec)

mysql> USE DATAGURU;
Database changed
mysql> SHOW TABLES;
+--------------------+
| Tables_in_dataguru |
+--------------------+
| career             |
| faculty            |
| grades             |
| newfaculty         |
| students           |
+--------------------+
5 rows in set (0.01 sec)

mysql>
```

SUMMARY

In this chapter you learned:

- **Importing**, **manipulating**, and **exporting** data, the three major procedures to data management.

- **Numeric** data types are used to store quantifiable values represented by numbers.

- The commands used to define columns that contain integers are **TINYINT**, **SMALLINT**, **MEDIUMINT**, **INT**, and **BIGINT**.

- **DECIMAL(*precision, scale*)** command is used to define data types which contain decimals.

- A **character** is a letter, digit, or special character, that is used to form a word or used as a separator. A sequence of characters is referred to as a **string**.

- **VARCHAR(*length*)** and **CHAR(*length*)** are used to define string data types, and they store data differently.

- There are five MySQL commands used to define dates and times: **DATE**, **TIME**, **DATETIME**, **TIMESTAMP**, and **YEAR**.

- The **INSERT INTO** command allows users to add rows to both empty tables or tables which already exist.

- The **LOAD DATA INFILE** command allows users to import data from a file into an existing file or into an empty table

“

If you don't know your numbers...

You don't know your business

MARCUS LEMONIS
THE PROFIT ON CNBC

DAY 3

In this chapter, you will:

- Learn the importance of data in business success
- Understand how to run basic queries to create reports and manipulate data
- Filter and sort data
- Save reports in a new table to be viewed later
- Export reports to a CSV file

DAY 3 OVERVIEW

The intent of Day 3 lessons is to get you started with manipulating data for analysis and decision-making. Today you'll begin extracting information from the **DATAGURU** database to simulate a real-world business scenario. You will complete the following tasks:

1) Run basic queries to explore all columns in a table
2) Run basic queries to explore subsets of columns in a table
3) Extract unique values from columns in a table
4) Apply a filter to obtain subsets of rows in a table
5) Sort data rows for readability
6) Create a new table as a subset from an existing table
7) Export data to a CSV file

In the last lesson, you installed the **DATAGURU** database. Take some time to explore the different tables in the database using previous MySQL commands you have learned. Be sure that you do not delete the databases, tables, columns, or rows.

GURU TIP

Learning to code is one of the most importat steps you can take to having a successful career in tech. Translating what you do into business value for clients and stakeholders is just as important.

BUSINESS INTELLIGENCE & ANALYTICS

Business Intelligence & Analytics (BI & A) is the use of business-related information to achieve company objectives. This information is located in databases and must be extracted and presented in a useful format to provide quick insights that help leaders successfully run their organizations. The action of requesting or extracting this information is call **querying**.

Database management and analytics are the lifeblood of an organization. Many organizations rely on these tools to identify opportunities to increase revenue, to strategically plan for the future, and to maintain daily business operations. Without insight into data, business leaders are making a guess on how to run their companies. In 2010, survey results published in Bloomberg Newsweek revealed that companies generating over $100 million in revenue used some form of BI&A.

The successful use of data can provide companies with a competitive advantage by assisting with the identification of when, who, and how to obtain and retain customers, learn more about competitors, and become more intelligible about business strategies.

So how do these decision-makers obtain a glimpse of this data? They regularly request & review reports. **Reports** are documents that contain organized summaries of information. In this lesson, you will learn how to query tables to create basic reports like those used in companies for strategic decision-making.

BASIC QUERIES

To create a database query a specific set of keywords are used. There are two statements which will always be used when querying data from a table in MySQL: **SELECT** and **FROM**. These two statements tell the computer what data to pull and where to obtain it from in the database. There are other commands which exist for data manipulation such as filter, sorting, and aggregating which can be used on an as needed basis.

The **SELECT** keyword statement is used to specify which columns the user wishes to view from a table. The user can select one or more columns from a table. The asterisk symbol is used to represent all columns. If the user wants to view all columns in a table they would type the command:

SELECT * FROM TABLE;

The **DISTINCT** keyword is an example of an optional command used for eliminating duplicate values in tables and reports. If the user desires to find unique values in a column then the *DISTINCT* keyword is placed after the *SELECT* statement.

SELECT DISTINCT * FROM TABLE;

The **FROM** keyword tells the computer which table the columns are located in. At some point, you might have to extract data from multiple tables simultaneously. The *FROM* statement allows this advanced capability which will be covered in Day 6 and Day 7.

BASIC QUERY EXAMPLE

In the TEST table shown below, there are five columns. You can select a subset of the columns by specifying the column names after the **SELECT** keyword. The order in which you designate the columns is the order in which the output will appear. On the next page, view the query to obtain only the **RowID**, **FavoriteFood**, and **LuckyNumber** columns from the TEST table.

TEST				
RowID	Name	BirthDate	FavoriteFood	LuckyNumber
000001	Jade	1995-09-24	Apple Pie	27
000002	Elliot	2001-02-22	Dried Squid	95
000003	Erin	1981-11-09	Tuna Pasta	22
000004	Alyce	1984-07-20	Adodo	35
000005	Nobuko	1984-09-11	Hamburger	19
000006	Yee	1985-07-20	Pizza	23
000007	Chante	1985-11-09	Barbeque	60
000008	Vivian	1986-10-09	Tacos	28
000009	Branda	1987-02-11	Banana Split	11
000010	Gisel	1987-05-13	Fried Chicken	5

BASIC QUERY EXAMPLE

```
SELECT
    FavoriteFood,
    RowID,
    LuckyNumber
FROM TEST;
```

FavoriteFood	RowID	LuckyNumber
Apple Pie	000001	27
Dried Squid	000002	95
Tuna Pasta	000003	22
Adodo	000004	35
Hamburger	000005	19
Pizza	000006	23
Barbeque	000007	60
Tacos	000008	28
Banana Split	000009	11
Fried Chicken	000010	5

FILTERING

When requesting data or creating reports, you can view a subset of rows that meet a specific condition or criteria. **Filtering** allows the user to extract a portion of data in which specified values meet a certain condition. A **condition** is a statement that can be either true or false.

The **WHERE** keyword is used to add filtering logic to database queries. Only data which is true for the condition shown in the *WHERE* clause will be returned in the query. In programming, conditions are defined using mathematical operators, as indicated below.

OPERATORS	SYMBOL	OPERATORS	SYMBOL
Addition	+	Greater Than Or Equal To	>=
Subtraction	-	Less Than	<
Multiplication	*	Less Than Or Equal To	<=
Division	/	Not Equal	<>
Exponentiation	^	Equal To	=
Modulo	%	And	AND
Greater Than	>	Or	OR

"WHERE" STATEMENT EXAMPLE

In the **TEST** table shown below, there are 10 rows. An option that can be used to select a subset of rows is the **WHERE** keyword. The *WHERE* keyword filters for specific values. When this is done, the computer checks each row to identify the data values that meet the condition. On the next page, view the query to obtain only the rows which have a **LuckyNumber** value less than 25.

TEST				
RowID	Name	BirthDate	FavoriteFood	LuckyNumber
000001	Jade	1995-09-24	Apple Pie	27
000002	Elliot	2001-02-22	Dried Squid	95
000003	Erin	1981-11-09	Tuna Pasta	22
000004	Alyce	1984-07-20	Adodo	35
000005	Nobuko	1984-09-11	Hamburger	19
000006	Yee	1985-07-20	Pizza	23
000007	Chante	1985-11-09	Barbeque	60
000008	Vivian	1986-10-09	Tacos	28
000009	Branda	1987-02-11	Banana Split	11
000010	Gisel	1987-05-13	Fried Chicken	5

"WHERE" STATEMENT EXAMPLE

```
SELECT
    FavoriteFood,
    RowID,
    LuckyNumber
FROM TEST
WHERE LuckyNumber < 25;
```

FavoriteFood	RowID	LuckyNumber
Tuna Pasta	000003	22
Hamburger	000005	19
Pizza	000006	23
Banana Split	000009	11
Fried Chicken	000010	5

SORTING

When requesting data or creating reports, you can organize the results so that rows in the table are sequenced as needed. **Sorting** is a way to arrange rows in a sequence based on a specified criteria.

The **ORDER BY** keyword allows the user to rearrange a dataset based on the values of one or more columns. The *ORDER BY* keyword is followed by one or more column names. Each row is rearranged in alphabetical, numerical, or chronological order depending on the data type specified in the column.

The default sorting order is ascending; for dates is from earliest to latest, for strings is from A to Z, and for numeric values is from least to greatest. You can specify that you want to sort the table in descending order by using the keywords **DESCENDING** OR **DESC** after the column name.

If you wish to create a query to sort your table by multiple columns, you can do this by specifying each column after the *ORDER BY* keyword with comma separators between each column name.

"ORDER BY" EXAMPLE

The *ORDER BY* keywords allow the user to sort data. The data is sorted in ascending order by default. On the next page, view the query to sort the filtered **TEST** table by the **LuckyNumber**. You will see the numbers sorted from lowest at the top to highest on the bottom.

TEST					Sort Order ↓
RowID	Name	BirthDate	FavoriteFood	LuckyNumber	
000001	Jade	1995-09-24	Apple Pie	27	
000002	Elliot	2001-02-22	Dried Squid	95	
000003	Erin	1981-11-09	Tuna Pasta	22	4
000004	Alyce	1984-07-20	Adodo	35	
000005	Nobuko	1984-09-11	Hamburger	19	3
000006	Yee	1985-07-20	Pizza	23	5
000007	Chante	1985-11-09	Barbeque	60	
000008	Vivian	1986-10-09	Tacos	28	
000009	Branda	1987-02-11	Banana Split	11	2
000010	Gisel	1987-05-13	Fried Chicken	5	1

```sql
SELECT
    FavoriteFood,
    RowID,
    LuckyNumber
FROM TEST
WHERE LuckyNumber < 25;
ORDER BY LuckyNumber;
```

FavoriteFood	RowID	LuckyNumber
Fried Chicken	000010	5
Banana Split	000009	11
Hamburger	000005	19
Tuna Pasta	000003	22
Pizza	000006	23

"CREATE TABLE" EXAMPLE

In the event you would like to store query output in a new table for later use, the **CREATE TABLE** keywords will allow you to do so. Before your *SELECT* query, simply type *CREATE TABLE*, followed by the new table name of your choosing, the *AS* keyword, and lastly your query. This will save the output data in a new table which can be referenced at a later time even after your MySQL session is over. If a database is not specified, the table will be stored in the database the developer is currently utilizing. The example below shows how to store the output from our example query in a new table called **NEWTEST**.

```
CREATE TABLE NEWTEST AS
SELECT
    FavoriteFood,
    RowID,
    LuckyNumber
FROM TEST
WHERE LuckyNumber < 25;
ORDER BY LuckyNumber;
```

NEWTEST		
FavoriteFood	RowID	LuckyNumber
Fried Chicken	000010	5
Banana Split	000009	11
Hamburger	000005	19
Tuna Pasta	000003	22
Pizza	000006	23

INTERMEDIATE SELECT QUERY SYNTAX

```
-- Extract data from a table
<CREATE TABLE NEWTABLE AS>

SELECT
    Column-1
   <,...Column-n>

FROM DATABASE.TABLE

<WHERE condition(s)>

<ORDER BY Column-1,
          <...Column-n>>;
```

FINANCIAL AID

LESSON 3

QUERIES & REPORTS

FINANCIAL AID HELP

In this lesson, you will fulfill information requests that data analysts receive when working in the Business Intelligence department of a college. Imagine you are working for a local university. Their financial aid department recently received funding for a new scholarship, but only qualified students will be eligible for it. The financial aid director has sent you a series of data requests asking for reports showing which students qualify for the aid. Your task is to write queries to complete the requests.

EXERCISE 3.1

The financial aid director sends you an email informing you that all scholarship eligibility will be based on the students' **State** of residence. Use the instructions in the pseudocode below to obtain your answer.

```
/*
    Using the STUDENTS table
    Write a query to obtain  a unique list of states using the
        State column
*/
```

EXERCISE 3.2

After sending the results to the financial aid director, you receive a follow-up request. Only students from the state of Crystalmallow are eligible to receive the new scholarship. The financial aid department wants to notify those students of their prospective eligibility and needs their contact information. Use the examples in this chapter along with the instructions in the pseudocode below to write a query to obtain the needed data.

```
/*
    Using the STUDENTS table
    Select the FirstName, LastName, Email, and PhoneNumber
        columns from the table
    Use the WHERE statement to filter the table and return only
        columns where  State is equal to "Crystalmallow"
*/
```

EXERCISE 3.3

The financial aid department, is now allowing scholarship eligibility for three additional states: IceWheat, Newdell, and Shadowmoor. They request you to add students from the three states to the report. There are two ways to filter for multiple conditions on the same column. First, try this exercise using the **OR** statement along with the instructions in the pseudocode below to write a query to obtain the needed data.

```
/*
    Using the STUDENTS table
    Select the FirstName, LastName, Email, and PhoneNumber
        columns from the table
    Use the WHERE statement to filter and return only columns
        that meet the following condition:
            - State is equal to "Crystalmallow" or
            - State is equal to "Icewheat" or
            - State is equal to "Newdell" or
            - State is equal to "Shadowmoor"
*/
```

GURU TIP

Sometimes consistent capitilazation is not used when entering strings into tables, which can make it difficult to use them in filtering conditions. Use the **UPCASE()** or **LOWCASE()** function to get around this issue. Below are a few examples:

WHERE UPCASE(State) = "CRYSTALMALLOW"
WHERE LOWCASE(State) = "crystalmallow"

EXERCISE 3.4

The goal of this exercise is to obtain the same information using the **IN** statement instead of multiple **OR** condition statements. The *IN* statement allows the user to create a list of elements to search for matching values. If the element is in the list, then the condition is true. Use the *IN* statement along with the pseudocode below to write a query to obtain the needed data.

```
/*
  Using the STUDENTS table
  Select the FirstName, LastName, Email, and PhoneNumber
     columns from the table
  Use the WHERE statement to filter and return only columns
     where State is in a list containing the states
        ("Crystalmallow", "Icewheat", "Newdell", "Shadowmoor")
*/
```

EXRCISE 3.5

Before sending the report, you notice it looks a bit unorganized. To help your business partners in the financial aid department, take time to sort the the student names and locations for better readability. Use the query you wrote in the last exerecise and the pseudocode below for instructions.

```
/*
  Using the STUDENTS table
  Select each student's FirstName, LastName, Email,
     and PhoneNumber
  Use the WHERE statement to filter and return only columns
     where State is in a list containing the states
        ("Crystalmallow", "Icewheat", "Newdell", "Shadowmoor")
  Use the ORDER BY statement to sort the table by
     State, FirstName, and LastName columns
*/
```

EXERCISE 3.6

By now you've probably noticed that you've run the same query a few times to obtain the same report. Just in case you need this data again, let's save it in a new table using the ***CREATE TABLE AS*** command. Use the query you wrote in the last exercise and the pseudocode below for instructions.

```
/*
  Create a new table called FINAID_REPORT using the
     STUDENTS table
  Select each student's FirstName, LastName, Email,
     and PhoneNumber
  Use the WHERE statement to filter and return only columns
     where State is in a list containing the states
        ("Crystalmallow", "Icewheat", "Newdell", "Shadowmoor")
  Sort the table by columns State, FirstName, and LastName
*/
```

EXPORTING CSV FILES

Running queries and creating tables is done using a user interface typically only developers have access to. Usually business stakeholders will not be able to access these interfaces, nor will they want to, so *how do they gain access to data and reports*?

Exporting the data into a file or report is a common practice to share information with business partners, colleagues, and external parties without giving them access to the database and user interface. This is not only helpful for the data requester, but it is convenient for the developer and keeps other data secure from prying eyes.

In the next lesson, you will create a CSV file called **finaid_report.csv** which contains the information in the **FINAID_REPORT** table. To export the file, you must obtain the directory location on your computer where you desire to place the file. The directory and file name together will look similar to the format shown below:

'C:/temp/finaid_reportcsv'

The syntax and instructions to complete this task are shown on the following pages.

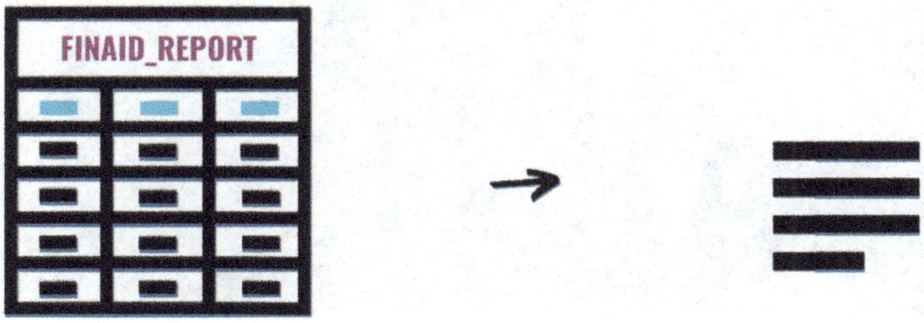

EXPORT FILE SYNTAX

```
-- Export data from a table into a file

SELECT
    Column-1
    <,...Column-n>

FROM TABLE

INTO OUTFILE '/directory/file'

FIELDS TERMINATED BY 'delimiter'

ENCLOSED BY string_identifier

LINES TERMINATED BY
        newline_identifier;
```

EXERCISE 3.7

Now that you have the data stored in a table, it's time to export it to a CSV file for submission to the financial aid office. Obtain the computer directory location of where file should be stored before you begin this lesson.

1) The first step is to tell the computer what data to put into the report. The data is in the FINAID_REPORT tables created in the last lesson. In the MySQL Command Line type **SELECT * FROM FINAID_REPORT** followed by pressing the Enter key.

2) Next, notify the computer of where to place the report by typing **INTO OUTFILE** followed by the file location and file name **finaid_report.csv** enclosed in single quotes. Then press the Enter key.

3) Since the output is a CSV file, indicate that the values in the file should be separated by commas by typing **FIELDS TERMINATED BY ','** then press the Enter key.

4) Next, specify that all string values in the file need to be enclosed in double quotes by typing **ENCLOSED BY '"'** and then press Enter.

5) Lastly, the computer needs to know how to separate each row of data. To tell the computer to put each row of data in the CSV file on a separate line, type **LINES TERMINATED BY '\n';** to indicate the newline character is where the row ends. Then press Enter.

Once all of the steps have been successfully completed, go to the file location on your computer to make sure your file exists. You will be able to view and share using Microsoft Excel. To get instructions on opening a CSV file in Excel visit csvexcel.dn2g.com

ANSWER 3.1

MySQL 8.0 Command Line Client

```
mysql> SELECT DISTINCT State FROM DATAGURU.STUDENTS;
+---------------+
| State         |
+---------------+
| Dorwall       |
| Crystalham    |
| Icewheat      |
| Aldmarble     |
| Marblemont    |
| Stoneelf      |
| Starryhollow  |
| Cordale       |
| Crystalmallow |
| Shadowmoor    |
| Newdell       |
| Butterloch    |
| Wheatmill     |
| Beechpond     |
| Southapple    |
| Shadowbell    |
```

ANSWER 3.2

MySQL 8.0 Command Line Client

```
mysql> SELECT FirstName,
    -> LastName,
    -> Email,
    -> MobileNumber
    -> FROM STUDENTS
    -> WHERE State = "Crystalmallow";
+-----------+----------+------------------------+----------------+
| FirstName | LastName | Email                  | MobileNumber   |
+-----------+----------+------------------------+----------------+
| Branda    | Galvan   | wsnyder@sbcglobal.net  | (953) 847-2269 |
| Lucinda   | Jacobs   | smallpaul@hotmail.com  | (406) 937-8776 |
| Alex      | Bartlett | augusto@icloud.com     | (789) 533-3435 |
| Junior    | Perry    | dbindel@live.com       | (900) 624-3214 |
+-----------+----------+------------------------+----------------+
4 rows in set (0.00 sec)

mysql>
mysql>
mysql>
mysql>
mysql>
mysql>
mysql>
mysql>
mysql>
mysql>
```

ANSWER 3.3

MySQL 8.0 Command Line Client

```
mysql> SELECT FirstName,
    -> LastName,
    -> Email,
    -> MobileNumber,
    -> State
    -> FROM STUDENTS
    -> WHERE State = "Crystalmallow" OR
    -> State = "Icewheat" OR
    -> State = "Newdell" OR
    -> State = "Shadowmoor";
+-----------+-----------+-------------------------+----------------+---------------+
| FirstName | LastName  | Email                   | MobileNumber   | State         |
+-----------+-----------+-------------------------+----------------+---------------+
| Eleonora  | Zuniga    | fudrucker@aol.com       | (707) 816-5715 | Icewheat      |
| Branda    | Galvan    | wsnyder@sbcglobal.net   | (953) 847-2269 | Crystalmallow |
| Gisela    | Howell    | heidrich@me.com         | (816) 453-0573 | Shadowmoor    |
| Mathilda  | Gordon    | mcraigw@att.net         | (617) 299-0334 | Newdell       |
| Yuko      | Monroe    | uraeus@aol.com          | (814) 418-4107 | Icewheat      |
| Lucinda   | Jacobs    | smallpaul@hotmail.com   | (406) 937-8776 | Crystalmallow |
| Nichol    | Case      | marioph@msn.com         | (703) 688-8284 | Shadowmoor    |
| Holly     | Rollins   | ghost@msn.com           | (314) 238-6499 | Newdell       |
| Rigoberto | Frederick | kohlis@optonline.net    | (545) 383-1696 | Icewheat      |
| Alex      | Bartlett  | augusto@icloud.com      | (789) 533-3435 | Crystalmallow |
| Ambrose   | Hamilton  | bebing@live.com         | (900) 345-6665 | Shadowmoor    |
| Allan     | Harper    | pajas@gmail.com         | (324) 294-2662 | Newdell       |
| Geoffrey  | Kerr      | mthurn@live.com         | (283) 501-0993 | Icewheat      |
| Junior    | Perry     | dbindel@live.com        | (900) 624-3214 | Crystalmallow |
| Winford   | Bush      | rafasgj@att.net         | (761) 965-8588 | Shadowmoor    |
| Micah     | West      | fraser@verizon.net      | (731) 555-3664 | Newdell       |
+-----------+-----------+-------------------------+----------------+---------------+
16 rows in set (0.00 sec)
```

You may have noticed, this is a very tedious way to filter on the same column.

ANSWER 3.4

MySQL 8.0 Command Line Client

```
mysql> SELECT FirstName,
    -> LastName,
    -> Email,
    -> MobileNumber,
    -> State
    -> FROM STUDENTS
    -> WHERE State IN ("Crystalmallow", "Icewheat", "Newdell", "Shadowmoor");
+-----------+-----------+-----------------------+----------------+---------------+
| FirstName | LastName  | Email                 | MobileNumber   | State         |
+-----------+-----------+-----------------------+----------------+---------------+
| Eleonora  | Zuniga    | fudrucker@aol.com     | (707) 816-5715 | Icewheat      |
| Branda    | Galvan    | wsnyder@sbcglobal.net | (953) 847-2269 | Crystalmallow |
| Gisela    | Howell    | heidrich@me.com       | (816) 453-0573 | Shadowmoor    |
| Mathilda  | Gordon    | mcraigw@att.net       | (617) 299-0334 | Newdell       |
| Yuko      | Monroe    | uraeus@aol.com        | (814) 418-4107 | Icewheat      |
| Lucinda   | Jacobs    | smallpaul@hotmail.com | (406) 937-8776 | Crystalmallow |
| Nichol    | Case      | marioph@msn.com       | (703) 688-8284 | Shadowmoor    |
| Holly     | Rollins   | ghost@msn.com         | (314) 238-6499 | Newdell       |
| Rigoberto | Frederick | kohlis@optonline.net  | (545) 383-1696 | Icewheat      |
| Alex      | Bartlett  | augusto@icloud.com    | (789) 533-3435 | Crystalmallow |
| Ambrose   | Hamilton  | bebing@live.com       | (900) 345-6665 | Shadowmoor    |
| Allan     | Harper    | pajas@gmail.com       | (324) 294-2662 | Newdell       |
| Geoffrey  | Kerr      | mthurn@live.com       | (283) 501-0993 | Icewheat      |
| Junior    | Perry     | dbindel@live.com      | (900) 624-3214 | Crystalmallow |
| Winford   | Bush      | rafasgj@att.net       | (761) 965-8588 | Shadowmoor    |
| Micah     | West      | fraser@verizon.net    | (731) 555-3664 | Newdell       |
+-----------+-----------+-----------------------+----------------+---------------+
16 rows in set (0.04 sec)
```

ANSWER 3.5

MySQL 8.0 Command Line Client

```
mysql> SELECT FirstName,
    -> LastName,
    -> Email,
    -> MobileNumber,
    -> State
    -> FROM STUDENTS
    -> WHERE State IN ("Crystalmallow", "Icewheat", "Newdell", "Shadowmoor")
    -> ORDER BY State, FirstName, LastName;
+-----------+-----------+-----------------------+----------------+---------------+
| FirstName | LastName  | Email                 | MobileNumber   | State         |
+-----------+-----------+-----------------------+----------------+---------------+
| Alex      | Bartlett  | augusto@icloud.com    | (789) 533-3435 | Crystalmallow |
| Branda    | Galvan    | wsnyder@sbcglobal.net | (953) 847-2269 | Crystalmallow |
| Junior    | Perry     | dbindel@live.com      | (900) 624-3214 | Crystalmallow |
| Lucinda   | Jacobs    | smallpaul@hotmail.com | (406) 937-8776 | Crystalmallow |
| Eleonora  | Zuniga    | fudrucker@aol.com     | (707) 816-5715 | Icewheat      |
| Geoffrey  | Kerr      | mthurn@live.com       | (283) 501-0993 | Icewheat      |
| Rigoberto | Frederick | kohlis@optonline.net  | (545) 383-1696 | Icewheat      |
| Yuko      | Monroe    | uraeus@aol.com        | (814) 418-4107 | Icewheat      |
| Allan     | Harper    | pajas@gmail.com       | (324) 294-2662 | Newdell       |
| Holly     | Rollins   | ghost@msn.com         | (314) 238-6499 | Newdell       |
| Mathilda  | Gordon    | mcraigw@att.net       | (617) 299-0334 | Newdell       |
| Micah     | West      | fraser@verizon.net    | (731) 555-3664 | Newdell       |
| Ambrose   | Hamilton  | bebing@live.com       | (900) 345-6665 | Shadowmoor    |
| Gisela    | Howell    | heidrich@me.com       | (816) 453-0573 | Shadowmoor    |
| Nichol    | Case      | marioph@msn.com       | (703) 688-8284 | Shadowmoor    |
| Winford   | Bush      | rafasgj@att.net       | (761) 965-8588 | Shadowmoor    |
+-----------+-----------+-----------------------+----------------+---------------+
16 rows in set (0.00 sec)
```

Notice how all of the rows are sorted by name in ascending order and location. The report is easier to view and you can see the number of students residing in each state.

ANSWER 3.6

MySQL 8.0 Command Line Client

```
mysql> CREATE TABLE FINAID_REPORT AS
    -> SELECT FirstName,
    -> LastName,
    -> Email,
    -> MobileNumber,
    -> State
    -> FROM STUDENTS
    -> WHERE State IN ("Crystalmallow", "Icewheat", "Newdell", "Shadowmoor")
    -> ORDER BY State, FirstName, LastName;
Query OK, 16 rows affected (0.42 sec)
Records: 16  Duplicates: 0  Warnings: 0

mysql> SHOW TABLES;
+--------------------+
| Tables_in_dataguru |
+--------------------+
| career             |
| finaid_report      |
| grades             |
| students           |
+--------------------+
4 rows in set (0.00 sec)

mysql> SELECT * FROM FINAID_REPORT;
+-----------+----------+------------------------+----------------+---------------+
| FirstName | LastName | Email                  | MobileNumber   | State         |
+-----------+----------+------------------------+----------------+---------------+
| Alex      | Bartlett | augusto@icloud.com     | (789) 533-3435 | Crystalmallow |
| Branda    | Galvan   | wsnyder@sbcglobal.net  | (953) 847-2269 | Crystalmallow |
| Junior    | Perry    | dbindel@live.com       | (900) 624-3214 | Crystalmallow |
| Lucinda   | Jacobs   | smallpaul@hotmail.com  | (406) 937-8776 | Crystalmallow |
| Eleonora  | Zuniga   | fudrucker@aol.com      | (707) 816-5715 | Icewheat      |
```

You should see 6 rows of data representing the number of students in the history class.

ANSWER 3.7

MySQL 8.0 Command Line Client

```
mysql> SELECT * FROM FINAID_REPORT
    -> INTO OUTFILE 'C:/temp/finaid_report.csv'
    -> FIELDS TERMINATED BY ','
    -> ENCLOSED BY '"'
    -> LINES TERMINATED BY '\n';
Query OK, 16 rows affected (0.03 sec)

mysql>
```

ANSWER 3.7

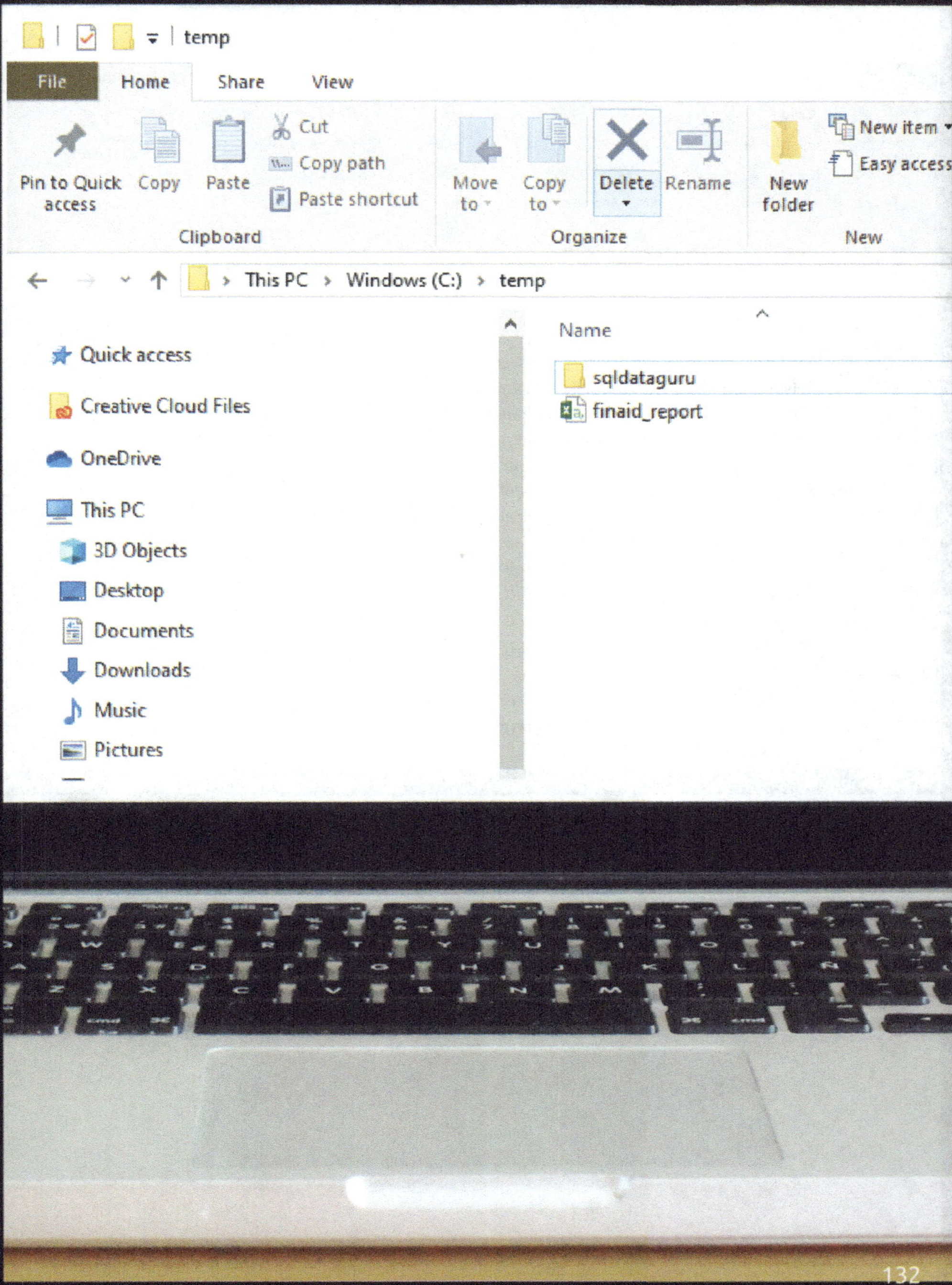

temp
File
Home
Share
View
Pin to Quick access
Copy
Paste
Cut
Copy path
Paste shortcut
Clipboard
Move to
Copy to
Delete
Rename
Organize
New folder
New item
Easy access
New
This PC › Windows (C:) › temp
Name
Quick access
Creative Cloud Files
OneDrive
This PC
3D Objects
Desktop
Documents
Downloads
Music
Pictures
sqldataguru
finaid_report

ANSWER 3.7

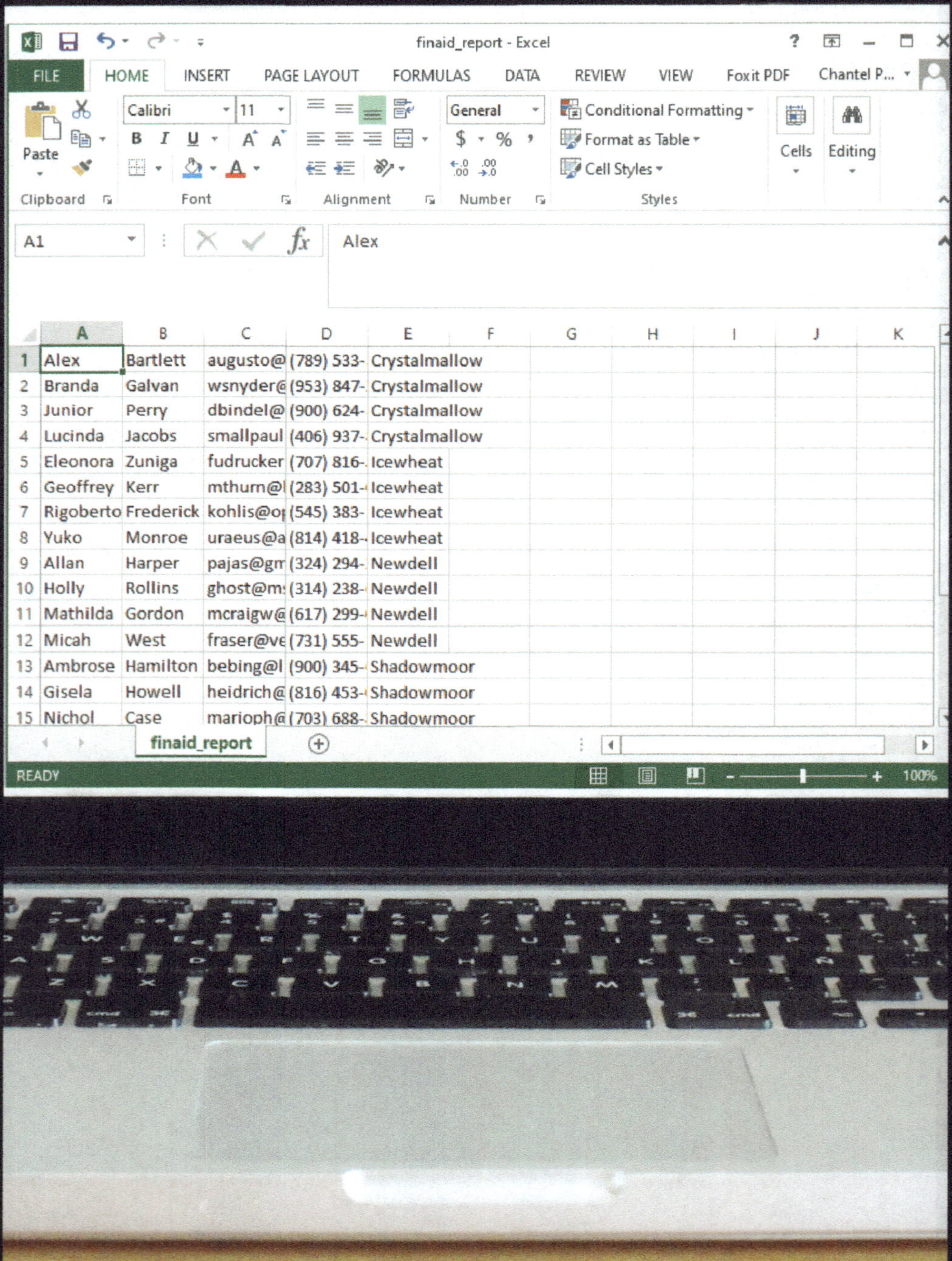
finaid_report - Excel
FILE HOME INSERT PAGE LAYOUT FORMULAS DATA REVIEW VIEW Foxit PDF Chantel P...
Paste
Calibri
11
General
Conditional Formatting
Format as Table
Cell Styles
Cells
Editing
Clipboard
Font
Alignment
Number
Styles
A1
Alex
1 Alex Bartlett augusto@ (789) 533- Crystalmallow
2 Branda Galvan wsnyder@ (953) 847- Crystalmallow
3 Junior Perry dbindel@ (900) 624- Crystalmallow
4 Lucinda Jacobs smallpaul (406) 937- Crystalmallow
5 Eleonora Zuniga fudrucker (707) 816- Icewheat
6 Geoffrey Kerr mthurn@ (283) 501- Icewheat
7 Rigoberto Frederick kohlis@o (545) 383- Icewheat
8 Yuko Monroe uraeus@a (814) 418- Icewheat
9 Allan Harper pajas@gm (324) 294- Newdell
10 Holly Rollins ghost@m (314) 238- Newdell
11 Mathilda Gordon mcraigw@ (617) 299- Newdell
12 Micah West fraser@ve (731) 555- Newdell
13 Ambrose Hamilton bebing@l (900) 345- Shadowmoor
14 Gisela Howell heidrich@ (816) 453- Shadowmoor
15 Nichol Case marioph@ (703) 688- Shadowmoor
finaid_report
READY
100%

SUMMARY

In this chapter you learned:

- Businesses analyze data to successfully plan and to operate their companies

- At the very least, a query to extract data from a table must have a **SELECT** and a **FROM** statement

- The **SELECT** keyword statement is used to specify which columns the user wishes to view from a table

- The **DISTINCT** keyword is an optional command used to eliminate duplicate values in tables and reports.

- The **FROM** keyword tells the computer which table the columns are located in.

- The **WHERE** keyword is used to add filtering logic to database queries.

- The **CREATE TABLE** keywords allow developers to store query output in a new table for later use.

- The **INTO OUTFILE** command allows you to export data from a an existing table into a file

“

If you don’t like something, change it.

If you can’t change it, change your attitude.

MAYA ANGELOU

DAY 4

In this chapter, you will:

- Learn the concept of data transformations
- Understand functions in programming
- Learn common functions used in MySQL
- Get familiar with CASE statements
- Create new data using functions

DAY 4 OVERVIEW

Congratulations, you made it to Day 4 or the halfway point in this book! Believe it or not, you have acquired the basic SQL skills needed to start building database storage systems similar to an ETL developer. In today's lesson you will be introduced to additional advanced querying concepts.

In previous chapters, you learned that data is presented in many forms and types. Data collected in its raw state is rarely sufficient or formatted adequately for clients to perform an analysis of the data. In this lesson, you will learn how to add new columns to datasets using two methods: (1) by transforming existing data and (2) by creating new data. You will complete the following tasks:

1) Create new columns using built-in functions
2) Create new columns using built-in functions and existing columns
3) Apply one condition CASE statements to transform existing data
4) Apply multi-condition CASE statements to transform existing data

You will use mathematical and boolean logic operators and functions to create new columns.

GURU TIP

Raw data is information which has not been cleaned or corrected for business use. Companies often have automated ETL jobs or scheduled processes which change the way the raw data is represented. The data is then stored in a secure location to be accessed for reporting and analysis.

DATA TRANSFORMATIONS

Data transformation is one of the most important phases in using data for analysis because it allows the user to address data quality issues by manipulating, subsetting, or even substituting data. If you use bad quality data for analysis, then your data will inherently be bad.

Occasionally, data may be missing from your dataset. One way to solve this issue is to impute or substitute data with other values. In analytics, substituting missing data values is commonly referred to as **data imputation**. In this chapter, you'll learn techniques you can use to impute missing or null values.

There are instances in which you may be required to change a column's data type. For example, when traveling from one country to another country that has different currency, you must convert money to make purchases. If you live in the United States and travel to Mexico, you will convert dollars to pesos because your currency is not in the correct form. The same thing can happen with data. For instance, you might have a column containing a number that represents the day of the week (i.e. 7), but you want to display the string version (i.e. Sunday). There are data transformation techniques to perform this task.

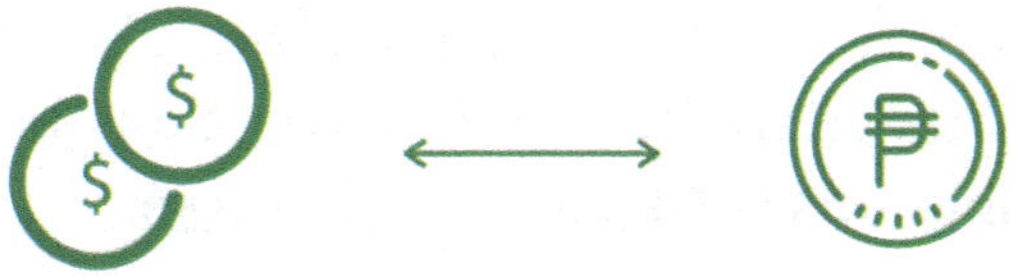

Lastly, a common data transformation technique is to use existing data to develop new data. For example, if you have a column containing a birth date, you can easily create a new column which contains the individual's age by subtracting the date of birth from the current date.

FUNCTIONS

Functions are a fundamental coding concept that all programming languages provide which allow the user to perform a routine process or specific task. Think of functions like a recipe. A recipe is a detailed set of instructions which assists chefs in completing the same task. To prepare the dish as instructed, the cook must use very specific ingredients and follow every step to obtain the desired outcome. Much like recipes most functions accept data values, process them using a series of steps, and output new values.

Each function has a unique name assigned to a specified task. You have seen functions throughout this book such as **UPCASE()** and **LOWCASE()**. These are MySQL built-in functions created by the MySQL development team which are available when using the software. You can also create your own functions, but this is a more advanced topic which will not be covered in this book. You can find out more about how to create your own functions, also known as **user-defined functions (UDF),** at udf.dn2g.com.

GURU TIP

In programming , functions may be referred to as procedures, subroutines, statements, methods, or subprograms.

MATHEMATICAL FUNCTIONS

Mathematical functions may be used to perform tasks, manipulate numeric data, represent complex numbers, or to depict mathematical symbols. When using a scientific calculator, there are a number of symbols that do not appear on a computer keyboard such as Π (pi) or √ (square root). In finance, engineering, and other math related fields these operators are often used for data analysis and reporting. There are functions which represent these symbols and perform corresponding tasks in MySQL and other programming languages.

The **SQRT()** function is used to calculate the √ (square root) of a nonnegative number. Let's look at an example of how this function is implemented on a column of values in MySQL. Given a column called Number located in a table called TABLE_EXAMP, the function will be used to create a new column called Sqrt_Number. In the figure below, the function is written after the *SELECT* statement and the Number column name is placed between two parentheses. The new column is assigned the name Sqrt_Number by specifying an alias with the *AS* keyword after the function is entered.

```
SELECT SQRT(Number) AS Sqrt_Number
FROM TABLE_EXAMP;
```

Functions can be used in other areas of a query where a column by itself would exists. For instance, it is possible to filter a data set to return only values with a square root of a number by implementing the function in a *WHERE* statement. The below example retains all rows where the square root of a Number is greater than 3.

```
SELECT *
FROM TABLE_EXAMP
WHERE SQRT(Number) > 3;
```

FUNCTION DIAGRAM

The illustration below depicts the process of a function using the **SQRT()** example from the previous page. Notice, each data element in the column is processed through the function and the output appears in a new column.

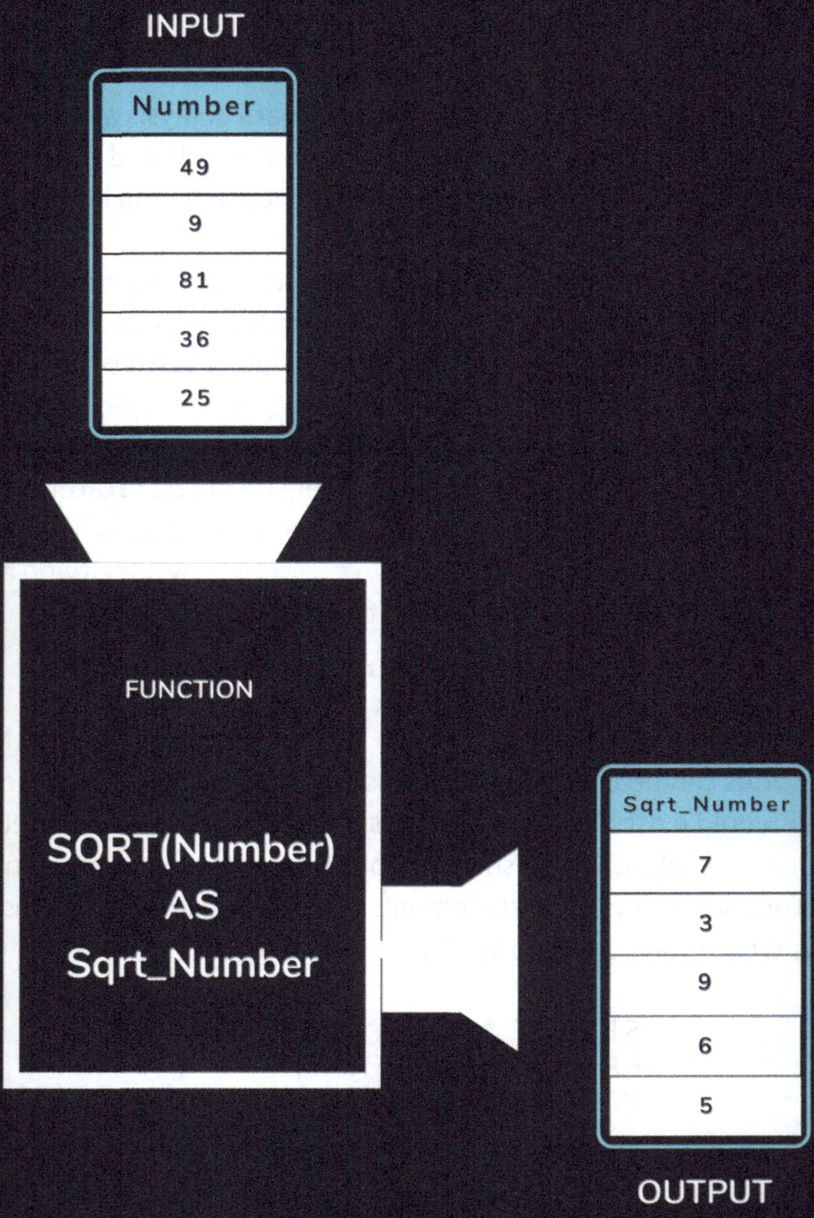

STRING FUNCTIONS

Functions don't only apply to numeric data types. They exist for string and date types as well. **String functions** allow the user to find, replace, or edit one or more strings at a time.

In the previous chapters, you worked with the **STUDENTS** table in various exercises. This table contains a **FirstName** and **LastName** column. It is common to see a person's first name and last name together from time to time (i.e. formal letters, mail, etc.) The example below, shows how the **CONCAT()** string function is used to connect two or more strings to create the column **ConcatName** which shows the first and last name together.

FirstName	LastName
Andrea	Nicole

Function		ConcatName
CONCAT(FirstName," ", LastName)	→	Andrea Nicole

Just as there are functions which link strings, there are also functions which separate strings. In the table below, there are a few different functions which return the same result. The **LEFT()** function retrieves a specified number of characters from the left side of a string column while the **RIGHT()** function retrieves the specified number of charactes from the right sideof a string column.

ConcatName	Function	Output
Andrea Nicole	LEFT(ConcatName, 6)	Andrea
	RIGHT(ConcatName, 6)	Nicole

DATE & TIME FUNCTIONS

Date and time functions are used to represent or process temporal values. These functions are helpful in calculating the time between events, tracking the occurrence of events, and extracting information from time values dates such as retrieving the month (May) or weekday (13th) from a timestamp ('05-13-2020 08:14:12'). One of the most common use cases for functions is calculating a person's age given their birth date.

Input	Function	Output
"2020-05-13"	MONTH("2020-05-13")	5
	DAY("2020-05-13")	13

DateOfBirth	Function
1998-02-17	TIMESTAMPDIFF(YEAR, DateofBirth, CURDATE())

Age
22

Not all functions require an input to produce an output like the **CURDATE()** and **CURTIME()** functions. Calling or implementing **CURDATE()** outputs the current date and calling **CURTIME()** returns the current time. These two functions are valuable for storing a record when changes are made and other reasons.

Function	Output
CURDATE()	2020-05-13
CURTIME()	08:14:12

On the following pages are common functions with corresponding examples. The MySQL team will add and remove functions over time when updating the software. Please refer to fxns.dn2g.com for the most recent list.

MATH FUNCTIONS

Name	Description	Example	Output
%, MOD	Modulo operator	16 % 4 19 mod 7	0 5
-	Change the sign of the argument	- 3	-3
ABS()	Return the absolute value	ABS(-7)	7
EXP()	Raise to the power of	EXP(2)	7.3890560
LOG()	Return the natural logarithm of the first argument	LOG(4)	1.3862943
PI()	Return the value of pi	PI()	3.141593
POWER()	Return the argument raised to the specified power	POWER(3,3)	27
ROUND()	Round the argument	ROUND(133.498,2)	133.50
SIGN()	Return the sign of the argument	SIGN(200) SIGN(-200)	1 -1
SQRT()	Return the square root of the argument	SQRT(49)	7

GURU TIP

Upgrade your software regularly to ensure you are obtaining new built-in functions offered by the MySQL development team.

STRING FUNCTIONS

Name	Description	Example	Output
CHAR_ LENGTH()	Return number of characters in argument	CHAR_LENGTH ("Dataguru")	8
CONCAT()	Return concatenated string	CONCAT("Push ","the ","door.")	Push the door.
INSERT()	Insert substring at specified position up to specified number of characters	INSERT("Hate it",1,4,"Love")	Love it
LEFT()	Return the leftmost number of characters as specified	LEFT("Dataguru",4)	Data
LENGTH()	Return the length of a string in bytes	LENGTH("United States")	13
LOCATE()	Return the position of the first occurrence of substring	LOCATE("8", "MySQL8.0")	6
LOWER()	Return the argument in lowercase	LOWER("DATAGURU")	dataguru

STRING FUNCTIONS

Name	Description	Example	Output
MID()	Return a substring starting from the specified position	MID("MySQL", 3,3)	SQL
REPEAT()	Repeat a string the specified number of times	REPEAT("Data", 2)	DataData
REPLACE()	Replace occurrences of a specified string	REPLACE("Apple Shake", "Apple", "Mango")	Mango Shake
REVERSE()	Reverse the characters in a string	REVERSE("Dataguru")	urugataD
RIGHT()	Return the specified rightmost number of characters	RIGHT ("Dataguru",4)	guru
SUBSTRING()	Return the substring as specified	SUBSTRING ("Data Transformation ",6,9)	Transform
UPPER()	Convert to uppercase	UPPER ("dataguru")	DATAGURU

DATE AND TIME FUNCTIONS

Name	Description	Example	Output
CURDATE()	Return the current date	CURDATE()	2020-04-19
CURTIME()	Return the current time	CURTIME()	17:31:55
DATEDIFF()	Subtract two dates	DATEDIFF("2020-04-20", "2020-04-15")	5
DAY()	Synonym for DAYOFMONTH()	DAY("2020-04-19")	19
DAYNAME()	Return the name of the weekday	DAYNAME("2020-04-19")	Sunday
DAYOF MONTH()	Return the day of the month (0-31)	DAYOFMONT H("2020-04-19")	19
DAYOF WEEK()	Return the weekday index of the argument	DAYOFWEEK("2020-04-19")	1
DAYOF YEAR()	Return the day of the year (1-366)	DAYOFYEAR("2020-04-19")	110

DATE AND TIME FUNCTIONS

Name	Description	Example	Output
HOUR()	Extract the hour	HOUR("2020-04-19 15:34:00")	15
MINUTE()	Return the minute from the argument	MINUTE("14:15:59")	15
MONTH()	Return the month from the date passed	MONTH("2020-04-19")	4
MONTH NAME()	Return the name of the month	MONTHNAME("2020-04-20")	April
STR_TO_DATE()	Convert a string to a date	STR_TO_DATE("APRIL 20 2020", "%M %d %Y")	2020-04-20
TIME()	Extract the time portion of the expression passed	TIME("2020-04-20 10:34:23")	10:34:23
YEAR()	Return the year	YEAR("2020-04-20")	2020

CONDITIONS

Conditions refer to one or more statements which can either be true or false. Depending on the outcome of the statement an action is taken. Below is a basic example of a condition along with a set of possible actions.

Condition
"Stacey ate breakfast this morning"

Actions when condition is true
"When Stacey eats breakfast in the morning, then she will eat a mid-morning snack"

Actions when condition is false
"Otherwise, Stacey will eat a mid-morning meal and skip lunch"

Conditions are used in programming when the computer must make a decision. The user creates the conditions and the possible set of outcomes or actions for the computer to take. In SQL, conditions are used within the **CASE** statement function to create new columns. *CASE* statements will be covered in the next section.

GURU TIP

In other coding languages, conditions are referred to as **if-then**, **if-then-else**, or a **switch** statements. They use different syntax keywords from **CASE** statements, but produce the same results.

CONDITION DIAGRAM

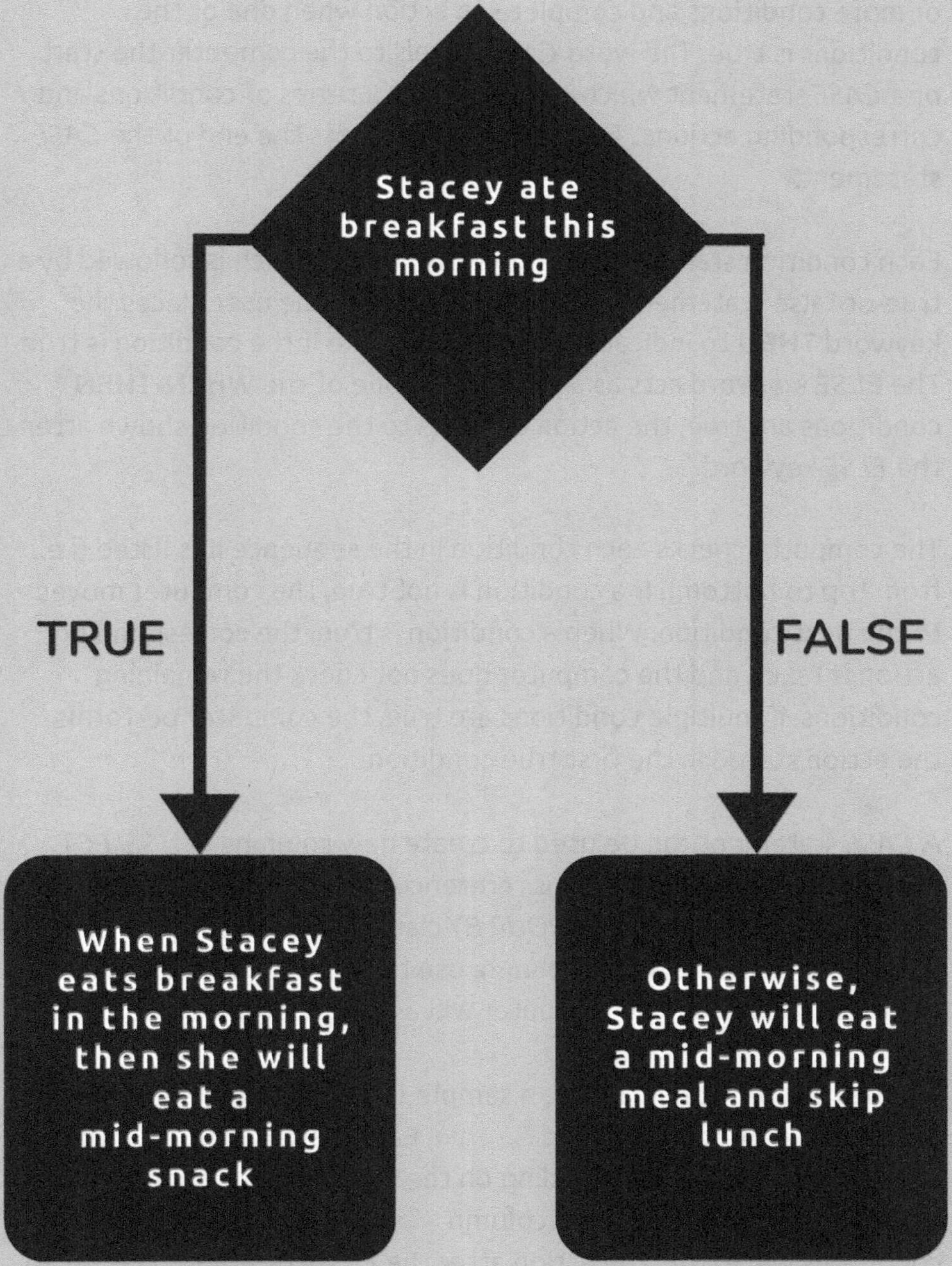

CASE STATEMENTS

CASE statements allow the computer to check through a list of one or more conditions and complete an action when one of the conditions is true. The word *CASE* signals to the computer the start of a *CASE* statement which is followed by a series of conditions and corresponding actions. The word **END** prompts the end of the *CASE* statement.

Each condition starts with the keyword **WHEN** which is followed by a true-or-false statement. After the statement, the user places the keyword **THEN** to indicates the action to take if the condition is true. The **ELSE** keyword acts as a catchall. If none of the **WHEN-THEN** conditions are true, the action defaults to the condition shown after the *ELSE* keyword.

The computer checks each condition in the sequence it is listed (i.e., from top to bottom). If a condition is not true, the computer moves to the next condition. When a condition is true, the corresponding action is taken and the computer does not check the remaining conditions. If multiple conditions are true, the computer performs the action stated in the first true condition.

A *CASE* statement can be used to create new columns in a *SELECT* query or anywhere a column is referenced in a query including after the *WHERE*, *ORDER BY*, or *GROUP BY* clauses. When using the *CASE* statement to create a new column, use the **AS** keyword followed by the column name or the computer will assign a default name.

On the following page, notice a sample *CASE* statement which is used to create a new column called **Region**. Each state is labeled with a corresponding region depending on the state value. A check is done on every row in the specified column **State**. Notice when none of the conditions were true, the action after the *ELSE* statement was taken (i.e. Region = 3).

CASE STATEMENT EXAMPLE

```
SELECT
StudentID,
(CASE
   WHEN State='Crystalmallow' THEN 1
   WHEN State='Shadowmoor' THEN 2
   WHEN State='Newdell' THEN 3
     ELSE 4
END) AS Region
FROM Students;
```

StudentID	State	Region
0000009	Crystalmallow	1
0000010	Shadowmoor	2
0000024	Icewheat	4
0000031	Shadowmoor	2
0000032	Newdell	3
0000053	Icewheat	4
0000059	Crystalmallow	1
0000060	Shadowmoor	2
0000061	Newdell	3
0000083	Crystalmallow	1

CASE STATEMENT SYNTAX

```
/* Check one or more conditions and when
one is true, then give the corresponding
outcome */

(
CASE
    WHEN condition(s)-1
    THEN outcome-1

    <WHEN condition(s)-n
        THEN outcome-n>

    ELSE outcome-m
END
)
```

Mar
Apr
May
Jun
Jul
Aug
Sep
Oct
Nov

TRANSFORM DATA

STUDENT OUTCOMES DILEMMA

In this exercise, a history professor noticed that the students might be having a rough time in her "HIST 396" class, but she isn't completely sure. She is curious to see if giving bonus assignments would help boost students scores to obtain a passing course grade. You will run a series of queries to help the teacher understand how her students are currently performing and if bonus assignments would help them pass the class.

EXERCISE 4.1

For the first data request, provide the history professor with a report of student's current scores using the pseudocode below:

```
/*
    Create a new table called HIST_Scores
    The table should include all columns from the Grades table
    Filter to obtain only the rows where "HIST 396" is the course
    The semester is "Fall" and the year is 2019
*/
```

How many students are in the history class?

EXERCISE 4.2

Now that the history class students grades have been obtained, it's time to figure out which students are passing or failing the course.

```
/*
Create a new table called HIST_STATUS
The table should include all columns from HIST_SCORES table
    Add a new column called Status which shows whether the
    student passed or failed the history course. Use a CASE
    statement   with the below conditions to create the column:
        - If the Score is less than 70, then 'FAIL'
        - If the Score is greater than or equal to 70, then 'PASS'
        - Otherwise leave it blank using two single quotes
*/
```

How many students are passing and how many are failing the history course?

EXERCISE 4.3

After sending the results to the professor, she sends a follow-up request. She is curious to know if bonus assignments might help her students. She is willing to offer assignments that will increase their current score by 35%. Write a query to see how an additional 35% bonus might enhance the students' scores.

```
/*
Create a new table called HIST_BONUS
The table should include all columns from HIST_STATUS table
Add a new column Bonus which shows the current Score with
        a 35% increase. Use the equation Score + (Score * .35) to
    create the new  column
*/
```

EXERCISE 4.4

Once the Bonus scores are retrieved, label them with a pass or fail status to see which students might be able to succeed after receiving a 35% increase.

```
/*
Create a new table called HIST_FINAL
The table should include all columns from HIST_BONUS table
Add a new column called BStatus which shows whether
    the student passed or failed the history course using the
    Bonus score.
   Use CASE statements with the following conditions to create
   the new column:
   - If Bonus is less than 70, then 'FAIL'
   - If the Bonus is greater than or equal to 70, then 'PASS'
   - Otherwise leave blank using space between two single quotes
*/
```

How many students would now receive a passing grade that had a failing grade before?

EXERCISE 4.5

In many American colleges, students receive a letter grade which represents how well they are mastering the course. The letter grades are often distributed to the school's provost to obtain an overall snapshot of student performance. In this exercise, you will create a report which provides letter grades by course subject. The course subject consists of the first four letters of the **Course** column which is located in the **GRADES** table.

```
/*
Run a query which outputs the following to the screen:
Use a string function to create a new column called Subject
    Create this column by obtaining only the first  4
    characters in the Course column which is located in the
    Grades table
Only display the first 10 rows to the screen
*/
```

There are various string functions which can be used to complete this task and return the same result. Can you find at least two different ways to obtain the same outcome?

EXERCISES 4.6

In the previous exercise, you created a **Subject** column using a string function. In this exercise, you will create a provost report which contains the **Subject** column along with the letter grades for each student. The conditions needed to create the letter grade are shown in the table and pseudo-code below.

A	90 - 100
B	80 - 89
C	70 - 79
D	60 - 69
E	50 - 59
F	0 - 49

```
/*
Create a new table called PROV_REPORT
The table should pull data from the Grades table and will only
     include  the 2 new columns below:
Add the new column called Subject that you created
     in Exercise 4.5
Add another column called LetterGrade which displays the
     letter which corresponds with the Score. Use a CASE
     statement  with the following conditions to create a column:
     - If the Score is >= 90 and <=100, then  "A"
        - If the Score is >= 80 and < 90, then  "B"
        - If the Score is >= 70 and < 80, then "C"
        - If the Score is >= 60 and < 70, then  "D"
        - If the Score is >= 50 and < 60, then "E"
        - If the Score is >= 0 and < 50, then  "F"
     - Otherwise leave blank using two double quotes
*/
```

There are a few ways to write the **CASE** statement. Can you find at least two different ways to obtain the same outcome?

ANSWER 4.1

MySQL 8.0 Command Line Client

```
mysql> USE DATAGURU;
Database changed
mysql> CREATE TABLE HIST_SCORES AS
    -> SELECT *
    -> FROM GRADES
    -> WHERE Course = "HIST 396"
    -> AND Semester = "Fall"
    -> AND Year = 2019;
Query OK, 6 rows affected (0.10 sec)
Records: 6  Duplicates: 0  Warnings: 0

mysql> SELECT * FROM HIST_SCORES;
+-----------+----------+------+----------+-------+
| StudentID | Semester | Year | Course   | Score |
+-----------+----------+------+----------+-------+
| 0000014   | Fall     | 2019 | HIST 396 |    76 |
| 0000036   | Fall     | 2019 | HIST 396 |    53 |
| 0000050   | Fall     | 2019 | HIST 396 |    67 |
| 0000053   | Fall     | 2019 | HIST 396 |    53 |
| 0000061   | Fall     | 2019 | HIST 396 |    50 |
| 0000078   | Fall     | 2019 | HIST 396 |    97 |
+-----------+----------+------+----------+-------+
6 rows in set (0.00 sec)

mysql>
```

You should see 6 rows of data representing the number of students in the history class.

ANSWER 4.2

MySQL 8.0 Command Line Client

```
mysql> CREATE TABLE HIST_STATUS AS
    ->  SELECT * ,
    -> (CASE
    -> WHEN Score < 70 THEN "FAIL"
    -> WHEN Score >= 70 THEN "PASS"
    -> ELSE ""
    -> END) AS Status
    -> FROM HIST_SCORES;
Query OK, 6 rows affected (0.11 sec)
Records: 6  Duplicates: 0  Warnings: 0

mysql> SELECT * FROM HIST_STATUS;
+-----------+----------+------+----------+-------+--------+
| StudentID | Semester | Year | Course   | Score | Status |
+-----------+----------+------+----------+-------+--------+
| 0000014   | Fall     | 2019 | HIST 396 |    76 | PASS   |
| 0000036   | Fall     | 2019 | HIST 396 |    53 | FAIL   |
| 0000050   | Fall     | 2019 | HIST 396 |    67 | FAIL   |
| 0000053   | Fall     | 2019 | HIST 396 |    53 | FAIL   |
| 0000061   | Fall     | 2019 | HIST 396 |    50 | FAIL   |
| 0000078   | Fall     | 2019 | HIST 396 |    97 | PASS   |
+-----------+----------+------+----------+-------+--------+
```

Notice that 4 out of 6 of the students are currently failing the course.

ANSWER 4.3

MySQL 8.0 Command Line Client

```
mysql> CREATE TABLE HIST_BONUS AS
    -> SELECT * ,
    -> Score + (Score *.35) AS Bonus
    -> FROM HIST_STATUS;
Query OK, 6 rows affected (0.04 sec)
Records: 6  Duplicates: 0  Warnings: 0

mysql> SELECT * FROM HIST_BONUS;
+-----------+----------+------+----------+-------+--------+--------+
| StudentID | Semester | Year | Course   | Score | Status | Bonus  |
+-----------+----------+------+----------+-------+--------+--------+
| 0000014   | Fall     | 2019 | HIST 396 |    76 | PASS   | 102.60 |
| 0000036   | Fall     | 2019 | HIST 396 |    53 | FAIL   |  71.55 |
| 0000050   | Fall     | 2019 | HIST 396 |    67 | FAIL   |  90.45 |
| 0000053   | Fall     | 2019 | HIST 396 |    53 | FAIL   |  71.55 |
| 0000061   | Fall     | 2019 | HIST 396 |    50 | FAIL   |  67.50 |
| 0000078   | Fall     | 2019 | HIST 396 |    97 | PASS   | 130.95 |
+-----------+----------+------+----------+-------+--------+--------+
6 rows in set (0.00 sec)
```

ANSWER 4.4

MySQL 8.0 Command Line Client

```
mysql> CREATE TABLE HIST_FINAL AS
    -> SELECT * ,
    -> (CASE
    -> WHEN Bonus < 70 THEN "FAIL"
    -> WHEN Bonus >= 70 THEN
    -> "PASS"
    -> ELSE ""
    -> END) AS BStatus
    -> FROM HIST_BONUS;
Query OK, 6 rows affected (0.18 sec)
Records: 6  Duplicates: 0  Warnings: 0

mysql> SELECT * FROM HIST_BONUS;
+-----------+----------+------+----------+-------+--------+--------+
| StudentID | Semester | Year | Course   | Score | Status | Bonus  |
+-----------+----------+------+----------+-------+--------+--------+
| 0000014   | Fall     | 2019 | HIST 396 |    76 | PASS   | 102.60 |
| 0000036   | Fall     | 2019 | HIST 396 |    53 | FAIL   |  71.55 |
| 0000050   | Fall     | 2019 | HIST 396 |    67 | FAIL   |  90.45 |
| 0000053   | Fall     | 2019 | HIST 396 |    53 | FAIL   |  71.55 |
| 0000061   | Fall     | 2019 | HIST 396 |    50 | FAIL   |  67.50 |
| 0000078   | Fall     | 2019 | HIST 396 |    97 | PASS   | 130.95 |
+-----------+----------+------+----------+-------+--------+--------+
6 rows in set (0.01 sec)
```

Two additional students would be able to pass the class if they successfully complete all of the bonus assignments.

ANSWER 4.5a

MySQL 8.0 Command Line Client

```
mysql> SELECT LEFT(Course, 4) AS Subject
    -> FROM GRADES
    -> LIMIT 10;
+---------+
| Subject |
+---------+
| MATH    |
| ENGR    |
| ENGR    |
| STAT    |
| HIST    |
| CHEM    |
| ENGR    |
| ENGR    |
| CHEM    |
| CHEM    |
+---------+
10 rows in set (0.07 sec)
```

ANSWER 4.5b

MySQL 8.0 Command Line Client

```
mysql> SELECT SUBSTRING(Course,1, 4) AS Subject
    -> FROM GRADES
    -> LIMIT 10;
+---------+
| Subject |
+---------+
| MATH    |
| ENGR    |
| ENGR    |
| STAT    |
| HIST    |
| CHEM    |
| ENGR    |
| ENGR    |
| CHEM    |
| CHEM    |
+---------+
10 rows in set (0.01 sec)
```

GURU TIP

Notice the condition, **LEFT(*col*, *num*)** is the same as **SUBSTRING(*col*, 1, *num*)** you will often find more than one way to do the same thing in programminng.

ANSWER 4.6a

MySQL 8.0 Command Line Client

```
mysql> CREATE TABLE PROV_REPORT AS
    -> SELECT SUBSTRING(Course,1, 4) AS Subject,
    -> (CASE
    ->   WHEN Score >= 90 AND Score <= 100 THEN "A"
    ->   WHEN Score >= 80 AND Score < 90 THEN "B"
    ->   WHEN Score >= 70 AND Score < 80 THEN "C"
    ->   WHEN Score >= 60 AND Score < 70 THEN "D"
    ->   WHEN Score >= 50 AND Score < 60 THEN "E"
    ->   WHEN Score >= 0 AND Score < 50 THEN "F"
    ->   ELSE ""
    ->  END) AS LetterGrade
    -> FROM Grades;
Query OK, 105 rows affected (0.10 sec)
Records: 105  Duplicates: 0  Warnings: 0

mysql>
```

ANSWER 4.6b

MySQL 8.0 Command Line Client

```
mysql> CREATE TABLE PROV_REPORT AS
    ->  SELECT SUBSTRING(Course,1, 4) AS Subject,
    ->  (CASE
    ->   WHEN Score BETWEEN 90 AND 100 THEN "A"
    ->   WHEN Score BETWEEN 80 AND 90 THEN "B"
    ->   WHEN Score BETWEEN 70 AND 80 THEN "C"
    ->   WHEN Score BETWEEN 60 AND 70 THEN "D"
    ->   WHEN Score BETWEEN 50 AND 60 THEN "E"
    ->   WHEN Score BETWEEN 0 AND 50 THEN "F"
    ->   ELSE ""
    ->  END) AS LetterGrade
    -> FROM Grades;
Query OK, 105 rows affected (0.12 sec)
Records: 105  Duplicates: 0  Warnings: 0

mysql>
```

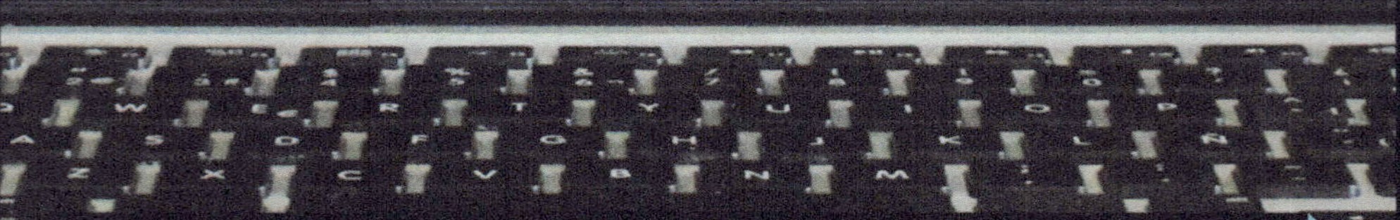

GURU TIP

Notice the condition, **>= X AND < Y** is the same as **BETWEEN X AND Y** and yields the same outcome.

SUMMARY

In this chapter you learned:

- **Data transformation** is an important phase in data analysis which is defined as any process in which the user manipulates, subsets, or substitutes data.

- **Data imputation** is the action of substituting missing data values.

- **Functions** are a fundamental coding concept that all programming languages provide which allow the user to perform a routine process or specific task.

- **Mathematical functions** may be used to perform tasks on manipulate numeric data, represent complex numbers, or to depict mathematical symbols.

- String functions allow the user to find, replace, or edit one or more strings at a time.

- **Date and time functions** are used to represent or process temporal values.

- **Conditions** refer to one or more statements which can either be true or false.

- **CASE statements** allow the computer to check through a list of one or more conditions and complete an action when one of the conditions is true.

"The people who are crazy enough to think they can change the world...

Are the ones who do.

STEVE JOBS

DAY 5

In this chapter, you will:

- Learn how to aggregate data to retrieve statistics
- Form groups of aggregates to investigate information in clusters
- Analyze data in your reports to retrieve insight
- Make business recommendations based off your analysis

DAY 5 OVERVIEW

When statistics are shown in the news or other media to quantify information, rarely are they presented with the underlying datasets and mathematical computations used to calculate the statistic. Typically, the media provides percentages, averages, or statistics which summarize a group of information. These **summary statistics** allow the public to quickly gain insight into a situation with one representative number.

Thus far, you've learned how to create basic to intermediate reports which focus on identifying individual rows of data. Today's lesson focuses on summarizing data from groups of rows in a table to provide insight. In this chapter, you will complete the tasks below:

1) Summarize all data in a table
2) Summarize data by category
3) Build summary reports
4) Provide insight from summary statistics

The goal of this lesson is to analyze data to find patterns in groups and provide a summary of your personal understanding of what these numbers mean. Keep in mind there are no rigid rules to what you infer and you may reach conclusions that aren't provided in this book. Explore the data and feel encouraged to consider solutions not presented in the answer key.

DATA GURU TIP

When analyzing data for business clients and partners, highlight key information that helps solve their problems. For example, a news headline and sub-headline are only one or two sentences. In-depth reports can be created for stakeholders who request further insights.

SUMMARY STATISTICS HEADLINES

Tesla Stock Rides Seven-Session Win Streak, Rising 56%

3 hours ago

By Alexander Osipovich

Video lottery operators still on hook for $7.5M bill despite shutdown

Amazon: 75,000 more jobs to fill due to coronavirus demand

Amazon has already filled 100,000 jobs over the last four weeks and needs an additional 75,000 workers to meet demand during the coronavirus crisis.

BUSINESS 4:05 p.m. ET Apr. 13

CLUSTERING

Clustering is a technique in data analysis used to detect patterns among groups. Data is split into groups that have similar characteristics and labeled for identification purposes. Each group is called a **cluster**.

After the groups are formed into clusters, analysts investigate them to find patterns in their behaviors. One of the easiest ways to understand clusters is to refer to the retail shopping industry. Retailers use analytics to cluster consumer information based on demographics such as age, income levels, ethnicity, location, etc.

Once retailers assemble the data, they examine shopper buying patterns and customize shopping experiences predicated on group preferences. For instance, Walmart is a large retail giant that utilizes customer geographic location to target their products. Shopping habits vary greatly for Walmart customers residing in New York City, NY compared to those in Miami, FL.

By analyzing customers purchases using location clusters, they can determine which items are in-demand in a particular region and provide the appropriate products. As a result, Walmart saves money by not stocking areas with products that are unpopular and they increase revenue by focusing their efforts on supplying items in locations where they will sell.

Clustering is a concept that is also used to understand diseases in healthcare, detect fraudulent activity at banks, track how people move in the travel industry and much more. In this section, you will learn how to analyze data in groups and make inferences about different clusters of students.

AGGREGATE FUNCTIONS

The last chapter covered numeric, string, date and time functions which allow the user to perform a task on individual rows in a column. **Aggregate functions** are a special type of functions which allow the user to look at summary data or statistics across a whole whole column or groups of rows.

The table below lists common aggregate functions used in MySQL along with their purpose. This table is not an all-inclusive list of summary statistics available in MySQL, but the full list can be accessed at aggr.dn2g.com. The next page shows the output of when using an aggregate function on a column of data.

Function	Description
COUNT()	Count the total number of values in a group
COUNT(DISTINCT)	Count the unique number of values in a group
AVG()	Average or mean of values in group
MIN()	Minimum value in group
MAX()	Maximum value in group
SUM()	Sum of values in group

AGGREGATE FUNCTIONS EXAMPLE

The Score column from the Grades table is shown on the left side of the diagram below. To obtain a statistics using an aggregate function place the function name after the *SELECT* statement as and place the column name to be aggregated between the parentheses. To rename the row simply add the *AS* keyword and the desired name behind the function. If a column name is not assigned a default name will be given as shown in the output below.

Grades
Score
95
67
89
89
63
62
84
62
54
52

```
SELECT MIN(Score)
FROM GRADES;
```

MIN(Score)
52

```
SELECT MAX(Score)
FROM GRADES;
```

MAX(Score)
95

```
SELECT SUM(Score)
FROM GRADES;
```

SUM(Score)
717

GROUP BY

The illustration on the previous page demonstrates how to summarize all data in a column, but *what if the user desires to aggregate data to view statistics across different groups or clusters?* The MySQL **GROUP BY** statement allows the user to obtain metrics across categories, such as:

Total Count of Students by State
Average Test Score by Student Gender
Youngest Student Age by Major

The *GROUP BY* statement is placed before the start of an *ORDER BY* clause and after any *WHERE* conditions in the query. The correct placement is shown on the next page. Make sure to state the column names for grouping both after the *SELECT* statement and the *GROUP BY* statement. Include the aggregate function in the *SELECT* statement as well.

After the *GROUP BY* statement, simply list the column names of one or more columns you wish to see aggregations grouped by. Additionally, programs aren't limited to grouping data to just one category. For instance, it's possible to obtain the highest student grades by course name and teacher.

When using multiple columns in a *GROUP BY* statement the order in which they are placed does not matter. The user is only required to separate each of the column names with a comma.

"GROUP BY" EXAMPLE #1

GRADES				
StudentID	Semester	Year	Course	StudentID
0000001	Summer	2018	MATH 120	95
0000006	Summer	2018	STAT 012	89
0000009	Summer	2018	CHEM 281	54
0000002	Spring	2019	ENGR 385	67
0000004	Spring	2019	ENGR 385	89
0000007	Spring	2019	HIST 396	63
0000007	Fall	2019	CHEM 281	62
0000008	Fall	2019	ENGR 385	84
0000009	Spring	2019	ENGR 385	62
0000013	Fall	2019	CHEM 281	52

This example uses a sample portion of data from the **GRADES** table, which contains student course scores, in the **DATAGURU** database. The highest score for each is calculated using the *GROUP BY* statement followed by the **Semester** column. Only the three unique semesters in the table are returned along with the corresponding maximum score. From the output table, one can infer "The highest grade for any Summer class was a score of 95" among other facts.

```
SELECT Semester,
MAX(Score) AS HiScore
FROM Grades
GROUP BY Semester;
```

Semester	HiScore
Summer	95
Spring	89
Fall	84

"GROUP BY" EXAMPLE #2

This example shows how a developer can obtain summary statistics across multiple columns. The highest score for each semester and course is calculated using the *GROUP BY* statement followed by the **Semester** and the **Course** column. Notice each column name is separated by a comma. Now there are seven unique groups in the table along with their corresponding maximum scores. From the output table, one can infer "The highest grade for any Summer MATH 120 class was a score of 95" among other facts.

```
SELECT Semester,
       Course,
       MAX(Score) AS HiScore
FROM Grades
GROUP BY Semester,
         Course;
```

Semester	Course	HiScore
Summer	MATH 120	95
Spring	ENGR 385	89
Summer	STAT 012	89
Spring	HIST 396	63
Fall	CHEM 281	62
Fall	ENGR 385	84
Summer	CHEM 281	54

HAVING

The example on the last page displays a table containing the highest grade by semester and year. What if you wanted to filter the table to obtain only the groups which had high scores greater than a certain number? In previous lessons, the *WHERE* statement was used to filter for specific rows. However, the *WHERE* statement does not work when filtering by group.

The **HAVING** clause is a special command used for filtering on groups of data. It is placed after the *GROUP BY* keyword, but before any *ORDER BY* statement if one exists. When the *HAVING* clause is specified, but no *GROUP BY* statement is present, the *HAVING* clause acts a regular *WHERE* statement because there are no groups present to filter on.

HAVING EXAMPLE

This scenario uses the **GRADES** table shown in the previous two examples. In the last example, the highest score for each semester and course was calculated. If the user desires to filter on these groups to find out which has a **HiScore** score greater than 70, then they may use the *HAVING* statement followed by the condition(s) on which they wish to filter.

```
SELECT Semester,
    Course,
    MAX(Score) AS HiScore
FROM Grades
GROUP BY Semester,
    Course
HAVING HiScore > 70;
```

Semester	Course	HiScore
Summer	MATH 120	95
Spring	ENGR 385	89
Summer	STAT 012	89
Fall	ENGR 385	84

ADVANCED SELECT QUERY SYNTAX

```
-- Extract data from a table
<CREATE TABLE NEWTABLE AS>

SELECT
    Column-1
    <,...Column-n>

FROM DATABASE.TABLE

<WHERE condition(s)>

<GROUP BY Column-1,
          <...Column-n>>

<ORDER BY Column-1,
          <...Column-n>>

<HAVING condition(s)>;
```

SUMMARIZE DATA

EARLY WARNINGS

Universities and colleges often have a Student Success department that performs analysis to identify students at risk of failing, stopping-out, or dropping out of school. In this lesson, you will assist the Student Success department in distinguishing student performance patterns by analyzing their grades. By examining grades across various groups you may find that one group tends to perform better or worse than another which would warrant further investigation and possible action to help that group of students.

EXERCISE 5.1

For your first task, you will use the aggregate functions to obtain statistics about all the data in the **Grades** table shown in the **DATAGURU** database. Follow the instructions in the pseudocode below to obtain the statistics.

```
/*
    Using data in the GRADES table
    Use the aggregate functions to create the following columns:
        - Create a column called CtScore which contains the count
            of Scores in the table
        - Create a column called CtID which contains the unique
            count of StudentID in the table
        - Create a MaxScore column which contains the Maximum
            Score in the table
        -Create a MinScore column which contains the Minimum
            Score in the table
        -Create a AvgScore column which contains the Average Score
            in the table
*/
```

EXERCISE 5.2

In this next exercise, you will investigate patterns by groups using the pseudo code below:

```
/*
    Using the query from the last exercise, cluster the statistics
        by Course. To cluster your data:
            - Include the Course column in your SELECT statement
            - Additionally, use a GROUP BY statement with the
            Course after the FROM keyword in your query
    At the end of the query, order the results by AvgScore
        from lowest to highest using the ORDER BY statement
*/
```

Which course groups seem to perform well (i.e. higher score statistics)? Which course groups seem to perform poorly (i.e. lower score statistics)? Use the clustered statistics to tell if there are certain groups performing above or below the group as a whole by comparing to the output form the last exercise?

EXERCISE 5.3

In this exercise you will compare the clustered group statistics to the total group statistics.

```
/*
    Using your query from the last exercise filter the clustered scores
    Filter out any clustered AvgScore which is less than the
        total AvgScore using the HAVING clause
    You can obtain the filter value for the total AvgScore by
        referring to the output from Exercise 5.1
*/
```

EXERCISE 5.4

When analyzing summary statistics of student performance in school, it is important to understand the range between their course scores. When the range between the high scores and low scores in the class is larger, this indicates some students are having widely different learning experiences in the classroom. On the other hand, if the range is smaller, then students have more similar score outcomes and learning experiences. In this exercise, you will view score ranges across different courses and make and inference one what this means for student learning.

```
/*
    Using data in the GRADES table
    Use the aggregate functions to create the following columns
        - Create a MaxScore column which contains the Maximum
        Score in the table
        - Create a MinScore column which contains the Minimum
        Score in the table
        - Create a DiffScore column which contains the
        difference between the Maximum and Minimum Score
         the table
    Next, cluster they data by Course
    Lastly, order the results by DiffScore from lowest to highest
        using the ORDER BY statement
*/
```

Do you notice scores which seem to have a larger difference in the minimum and maximum for certain courses? Are students performing more consistently in some courses than others?

EXERCISE 5.5

As you may have already noticed, when the summary statistics are broken into clusters the inference or conclusion about the statistic begins to change. In this next exercise, let's add another level of granularity by grouping statistics by Course and Semester.

```
/*
    Using your query from Exercise 5.2 cluster the statistics by
    Course and Semester
    To cluster your data include the Course and Semester column
    in your SELECT statement
    Additionally, use a GROUP BY statement along with the
    Course and Semester column after the FROM statement
    in your query
    Lastly, order the results by AvgScore from lowest to highest
    using the ORDER BY statement
*/
```

After viewing the new output, compare see if your previous thoughts on course performance. Is the course with the highest average scores the same in this exercise as the one in Exercise 5.2? Does adding the semester seem to impact the score outcomes? Regardless of the semester, are there any courses that seem to always produce consistently high or low average scores for students?

EXERCISE 5.6

In the last exercise, you viewed summary statistics by **Course** and **Semester**. Semesters group time frames up into various months and are one way to investigate timing patterns among groups. The **GRADES** table also contains the field **Year** which represents the year in which the course was offered and taken. In this exercise, you will analyze score statistics grouped by **Semester** and **Year**.

```
/*
    Using your query from Exercise 5.1 cluster the statistics
        by Semester and Year
    To cluster your data include the Semester and Year column in
        your SELECT statement
    Additionally, use a GROUP BY Semester and Year after the
        FROM statement in your query
    Lastly, order the results by AvgScore from lowest to highest
        using the ORDER BY statement
*/
```

Are there time periods where students perform better than others or do score statistics remain fairly constant over time?

ANSWER 5.1

MySQL 8.0 Command Line Client

```
mysql> SELECT COUNT(Score) AS CtScore,
    -> COUNT(Distinct StudentID) AS CtID,
    -> MAX(Score) AS MaxScore,
    -> MIN(Score) AS MinScore,
    -> AVG(Score) AS AvgScore
    -> FROM Grades;
+---------+------+----------+----------+----------+
| CtScore | CtID | MaxScore | MinScore | AvgScore |
+---------+------+----------+----------+----------+
|     101 |   64 |      100 |       50 |  72.8317 |
+---------+------+----------+----------+----------+
1 row in set (0.01 sec)

mysql>
```

ANSWER 5.2

MySQL 8.0 Command Line Client

```
mysql> SELECT Course,
    -> COUNT(Score) AS CtScore,
    -> COUNT(Distinct StudentID) AS CtID,
    -> MAX(Score) AS MaxScore,
    -> MIN(Score) AS MinScore,
    -> AVG(Score) AS AvgScore
    -> FROM Grades
    -> GROUP BY Course
    -> ORDER BY AvgScore ASC;
+----------+---------+------+----------+----------+----------+
| Course   | CtScore | CtID | MaxScore | MinScore | AvgScore |
+----------+---------+------+----------+----------+----------+
| STAT 012 |      18 |   17 |       91 |       50 |  66.3333 |
| ECON 204 |      12 |   12 |       83 |       51 |  67.3333 |
| HIST 396 |      12 |   12 |       97 |       50 |  68.5833 |
| CHEM 281 |      17 |   15 |       96 |       52 |  71.1176 |
| CS 193   |      14 |   12 |      100 |       50 |  76.5714 |
| ENGR 385 |      16 |   15 |       95 |       62 |  80.1875 |
| MATH 120 |      12 |   12 |       97 |       58 |  80.5833 |
+----------+---------+------+----------+----------+----------+
7 rows in set (0.00 sec)
```

Notice the "Math 120" and "ENGR 385" courses seem to perform the highest. On the other hand, "STAT 012" and "ECON 204" are at the lowest performing courses on average. You can use the MaxScore column to tell that at least one student in the group is performing above average and the MinScore to tell that a student is performing below average.

ANSWER 5.3

MySQL 8.0 Command Line Client

```
mysql> SELECT Course,
    -> COUNT(Score) AS CtScore,
    -> COUNT(Distinct StudentID) AS CtID,
    -> MAX(Score) AS MaxScore,
    -> MIN(Score) AS MinScore,
    -> AVG(Score) AS AvgScore
    -> FROM Grades
    -> GROUP BY Course
    -> HAVING AvgScore <=  72.813;
+----------+---------+------+----------+----------+----------+
| Course   | CtScore | CtID | MaxScore | MinScore | AvgScore |
+----------+---------+------+----------+----------+----------+
| CHEM 281 |      17 |   15 |       96 |       52 |  71.1176 |
| ECON 204 |      12 |   12 |       83 |       51 |  67.3333 |
| HIST 396 |      12 |   12 |       97 |       50 |  68.5833 |
| STAT 012 |      18 |   17 |       91 |       50 |  66.3333 |
+----------+---------+------+----------+----------+----------+
4 rows in set (0.09 sec)

mysql>
```

ANSWER 5.4

MySQL 8.0 Command Line Client

```
mysql> SELECT Course,
    -> MAX(Score) AS MaxScore,
    -> MIN(Score) AS MinScore,
    -> MAX(Score) - MIN(Score) AS DiffScore
    -> FROM Grades
    -> GROUP BY Course
    -> ORDER BY DiffScore;
+----------+----------+----------+-----------+
| Course   | MaxScore | MinScore | DiffScore |
+----------+----------+----------+-----------+
| ECON 204 |       83 |       51 |        32 |
| ENGR 385 |       95 |       62 |        33 |
| MATH 120 |       97 |       58 |        39 |
| STAT 012 |       91 |       50 |        41 |
| CHEM 281 |       96 |       52 |        44 |
| HIST 396 |       97 |       50 |        47 |
| CS 193   |      100 |       50 |        50 |
+----------+----------+----------+-----------+
7 rows in set (0.02 sec)
```

The "CS 193" course has the largest difference in highest and lowest scores, while "ECON 204" has the smallest difference which means their students tend to perform more similarly than students in the other courses.

ANSWER 5.5

MySQL 8.0 Command Line Client

```
mysql> SELECT Course,
    -> Semester,
    -> COUNT(Score) AS CtScore,
    -> COUNT(Distinct StudentID) AS CtID,
    -> MAX(Score) AS MaxScore,
    -> MIN(Score) AS MinScore,
    -> AVG(Score) AS AvgScore
    -> FROM Grades
    -> GROUP BY Course, Semester
    -> ORDER BY AvgScore ASC;
```

Course	Semester	CtScore	CtID	MaxScore	MinScore	Avg
ECON 204	Summer	1	1	55	55	55
MATH 120	Spring	1	1	58	58	58
CHEM 281	Summer	4	4	68	53	58
ECON 204	Fall	4	4	62	51	59
STAT 012	Spring	5	5	81	53	63
HIST 396	Fall	6	6	97	50	66
STAT 012	Fall	10	10	91	50	66

Even with Semester granularity, "MATH 120" remains at the top. However, "STAT 012" is no longer lagging at the bottom. None of the students in those courses seem to perform consistently higher or lower than the others when grouped by semester.

ANSWER 5.6

MySQL 8.0 Command Line Client

```
mysql> SELECT Semester,
    -> Year,
    -> COUNT(Score) AS CtScore,
    -> COUNT(Distinct StudentID) AS CtID,
    -> MAX(Score) AS MaxScore,
    -> MIN(Score) AS MinScore,
    -> AVG(Score) AS AvgScore
    -> FROM Grades
    -> GROUP BY  Semester, Year
    -> Order by AvgScore ASC;
+----------+------+---------+------+----------+----------+----------+
| Semester | Year | CtScore | CtID | MaxScore | MinScore | AvgScore |
+----------+------+---------+------+----------+----------+----------+
| Spring   | 2019 |      32 |   29 |       94 |       50 |  72.4688 |
| Fall     | 2019 |      50 |   42 |      100 |       50 |  72.8400 |
| Summer   | 2018 |      19 |   16 |       97 |       53 |  73.4211 |
+----------+------+---------+------+----------+----------+----------+
3 rows in set (0.09 sec)

mysql>
```

Each semester during 2018 and 2019 has a consistent average score around 72 or 73. Additionally, the minimum and maximum scores are close in value.

SUMMARY

In this chapter you learned:

- **Summary statistics** allow the for quick insight into underlying data with one representative number.

- **Clustering** is a technique in data analysis used for detecting patterns amongst groups. Data is formed into groups with similar characteristics called a **cluster**.

- **Aggregate functions** are a special group of functions which allow the user to look at summarized data or statistics.

- The MySQL **GROUP BY** statement allows the user to obtain metrics across one or more categories.

- When filtering for aggregate values use the **HAVING** clause because the **WHERE** statement does not work for groups.

- Adding more granularity or grouping by more categories may change your inference about a summary statistic.

“

Data are just summaries of thousands of stories –

Tell a few of those stories to help make the data meaningful

EDWARD TUFTE

DAY 6

In this chapter, you will:

- Learn different methods to collect data from various source
- Understand how to sketch designs to plan prior to combining data
- Create joins to combine data horizontally
- Use conditions to combine tables

DAY 6 OVERVIEW

Most data professionals work on complex projects in which solicited data is located in one or more tables. In these instances, data sources must be merged to obtain the required information.

In today's lesson, you will learn how to combine one or more tables using a **JOIN** statement. *JOIN* statements allow users to horizontally combine tables by row, based on one or more matching columns or other criteria. The following sections cover various types of table joins. You will complete the following tasks:

1) Plan and diagram table joins
2) Join two tables with the left side assigned as the master table
3) Join two tables with the right side assigned as the master table.

Today's lesson is more advanced and it may take time and practice to get comfortable using joins. So, hang in there!

You are only 2 lessons away from achieving Data Guru status!

GURU TIP

The task of joining can be a bit confusing and it is also computationally expensive, therefore, mapping is a critical skill utilized to save time and money . Designing a plan before development can eliminate time consuming mistakes. Get into the habit of creating a plan and allowing teammates to review your design before running queries to ensure the best solutions and utilization of resources.

HORIZONTAL TABLE MERGE

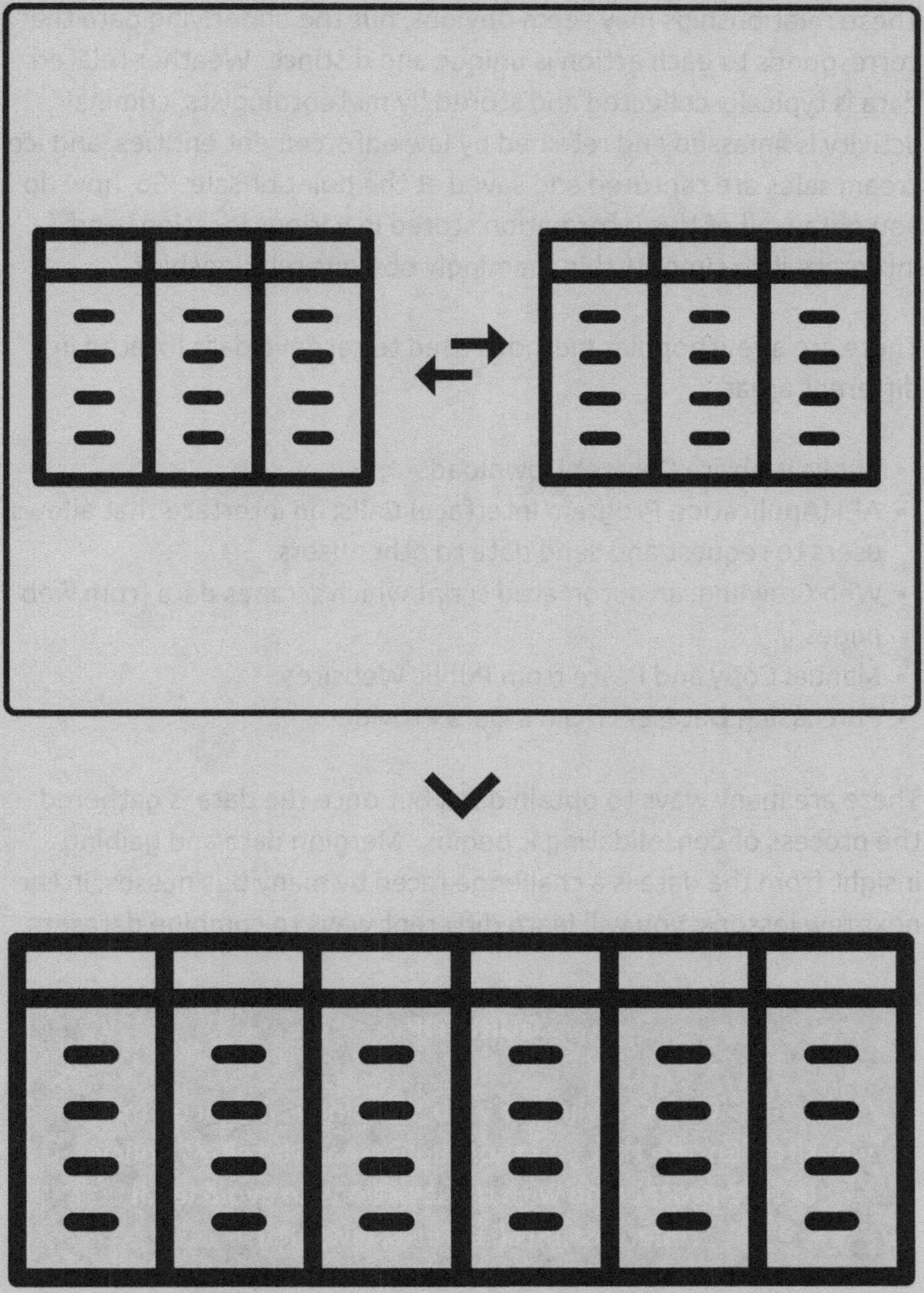

COLLECTING DATA

Did you know that an increase in outside temperatures typically correlates with an increase in violent crimes and ice cream sales? These relationships may seem obvious, but the underlying data that corresponds to each action is unique and distinct. Weather related data is typically collected and stored by meteorologists, criminal activity is amassed and retained by law enforcement entities, and ice cream sales are captured and saved at the point of sale. So, how do you obtain all of this information stored in various locations and integrate it to support this seemingly obvious relationship?

There are a few popular methods used to retrieve data located in different areas:

- Public Website Dataset Downloads
- API (Application Program Interface) Calls: an interface that allows users to request and send data to other users.
- Web Crawling: an automated script which scrapes data from web pages
- Manual Copy and Paste from Public Websites
- Purchasing Datasets from a Data Provider

There are many ways to obtain data, but once the data is gathered the process of consolidating it begins. Merging data and gaining insight from the data is a challenge faced by many businesses. In the next few lessons, you will learn different ways to combine datasets.

GURU TIP

WARNING Web Crawling or Web Scraping is illegal on many websites. Ensure that you thoroughly read the site's policies before proceeding with these actions to avoid legal repercussions (i.e. fines, lawsuits).

CRIME SCENE

TABLE JOINS

In the **DATAGURU** database, the **STUDENTS** table contains student email addresses while the **GRADES** table is composed of student grades. Since these tables are separate, how would you email the students a copy of their respective grades? The two tables can be merged to accomplish this task. In order to combine rows from different tables a **JOIN** statement is required.

There are two major types of joins referred to as **Outer** and **Inner**. The Outer join has three sub-types called **Full Outer**, **Left**, and **Right** joins. The Inner join do not have sub-types; it is singular. Today's lesson will cover Outer join types and tomorrow's lesson will include the concept of the Inner join.

In order to join two or more tables, they must have a common value or matching criteria. As stated in the Introductory chapter, RDMBS tables have a relationship with one another which means distinct data is stored in each table, but the tables correlate in some manner. Cardinality is used to explain the relationship between two or more datasets in a RDBMS. **Cardinality** refers to the uniqueness of data values in table columns. Tables that have columns with a high number of unique data values have higher cardinality. On the other hand, tables with repeating values have lower cardinality. The different relationship types are referred to as: **one-to-one, one-to-many, and many-to-many**.

Developers utilize one or more **keys,** or row identifiers to link tables. In a **one-to-one** relationship between tables, one row in one table is related to exactly one row in another table using a key column. In a **one-to-many** relationship, one row in a table is related to many rows in another table using a key column. In a **many-to-many** relationship, many rows in one table connect to many rows in another table using a key column. On the next page you are provided examples of the different cardinality types.

RELATIONSHIP TYPES

ONE-TO-ONE

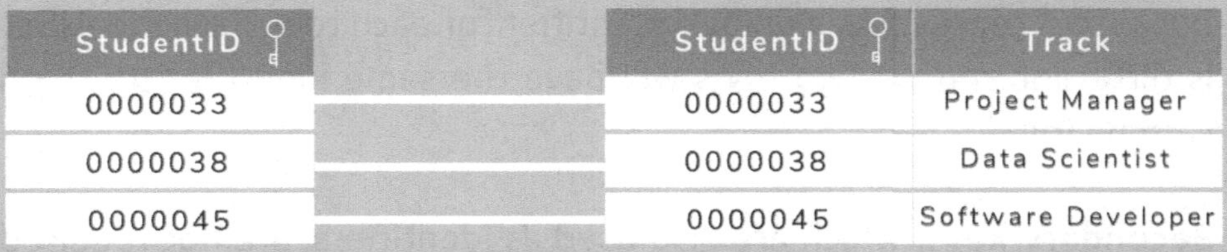

ONE-TO-MANY

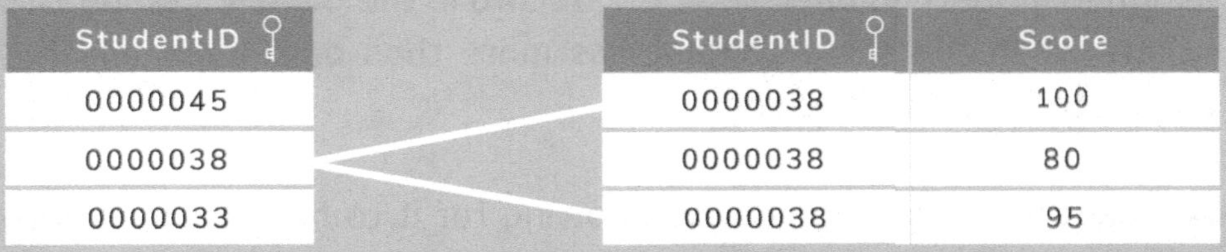

MANY-TO-MANY

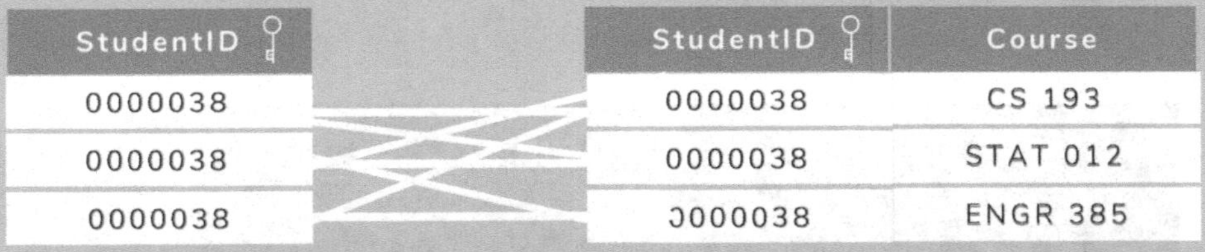

Note all key columns are labeled with a tiny key symbol next to the column name in this example.

JOIN KEYS

There are two types of relational database keys: **primary** and **secondary**. On the next page, notice a StudentID in the first column of both tables. Each student is assigned a unique Student ID that is used to identify and locate their information in each table and it also relates the tables to each other. This StudentID is a **primary key** because it represents a unique identifier for each row. Each identifier is different and no two rows will have the same identifier (i.e. high cardinality).

Secondary keys, which are also used as identifiers, are not required to be unique and may be shared among rows (i.e. low cardinality). The GRADES table also contains a StudentID, but in this case the identifier is a secondary key because the same student can be assigned more than one StudentID record in the GRADES table. This occurs when the same student has more than one grade and has completed multiple courses.

A table can have multiple keys allowing for it to have the ability to connect with various other tables. Each key within a table can only be primary or secondary. However, separate tables with the same key or identifier are not required to have the same classification. In a one-to-many relationship, the key is primary in one table and secondary in the other.

GURU TIP

Fuzzy joins are an informal join type defined as joining two or more tables that don't have an exact key match. When obtaining data from different data sources, fuzzy joins are a common occurrence. A developer may be able to obtain a close or approximate match using a "fuzzy join".

PRIMARY & SECONDARY KEY TABLE EXAMPLES

STUDENTS

StudentID	FirstName	MobileNumber
0000006	Yee	(344) 814-8732
0000007	Chantell	(204) 460-6256
0000008	Vivian	(965) 427-9466
0000009	Branda	(953) 847-2269
0000010	Gisela	(816) 453-0573

GRADES

StudentID	Course	Score
0000007	HIST 396	63
0000007	CHEM 281	62
0000008	ENGR 385	84
0000009	ENGR 385	62
0000009	CHEM 281	54

OUTER JOINS

A Venn diagram is used to depict possible relationships between two or more datasets. These diagrams show overlapping circles which represent the datasets and their relationships. In this section, Venn diagrams will be used to explain how table joins work. There are three types of Outer joins cited at the beginning of this chapter:

- **LEFT JOIN:** The left table is the master table in the join
- **RIGHT JOIN**: The right table is the master table in the join
- **FULL OUTER JOIN**: Both tables are equally important to the join and unmatched rows are returned from both tables.

When using a *LEFT JOIN*, the master or main table will be displayed on the left side in a Venn Diagram. Conversely, when using a *RIGHT JOIN* the master or main table will be shown on the right of the Venn Diagram. What does it mean to be the master table?

The master table is the dominant table which means:

1) When a join occurs, all data inside the master table is retained;
2) The partner table is attempting to correspond to the master;
3) If the keys from the partner table do not correspond to the master, those columns will display as blank (i.e. NULL) in the final table

On the following pages Venn diagrams are used to illustrate Outer joins using circles to represent Table A and Table B. The blue area signifies where data exists for the specified table. In a *LEFT JOIN* all data exists for Table A - because it is the master table located on the left side of the join - and data located in the center where the two tables intersect represent matching keys from Table B. The opposite is true for second illustration representing, the ***RIGHT JOIN***. Table B is now the master table since it is on the right of the join. The third illustration represents the *FULL OUTER JOIN* which includes all data regardless of whether the key matches or not because both tables are equally important.

OUTER JOIN VENN DIAGRAMS

LEFT JOIN

TABLE A is located on the left (master), and all data from Table A will be present when the tables are joined.

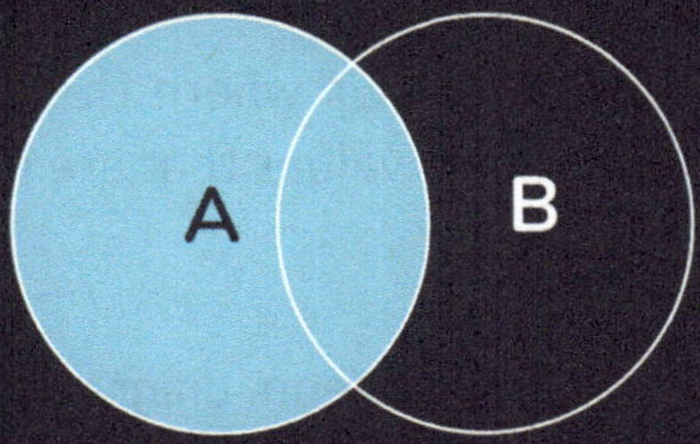

TABLE B is the partner table, and data in Table B will only be shown when the join keys match with Table A.

RIGHT JOIN

TABLE A is the partner table, and data in Table A will only be shown when the join keys match with Table B.

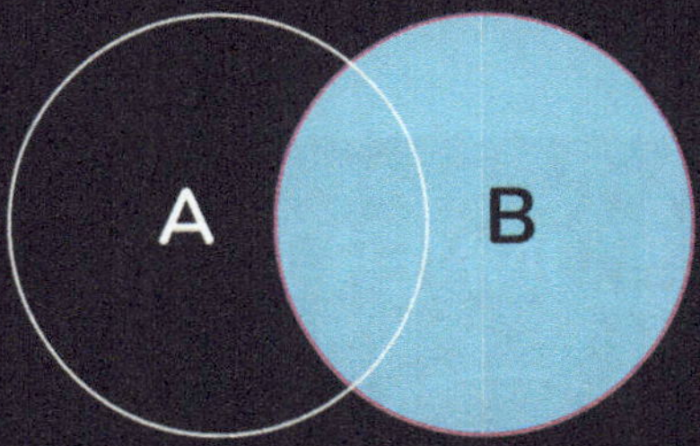

TABLE B is located on the right (master), all data from Table B will be present when the tables are joined.

FULL OUTER JOIN

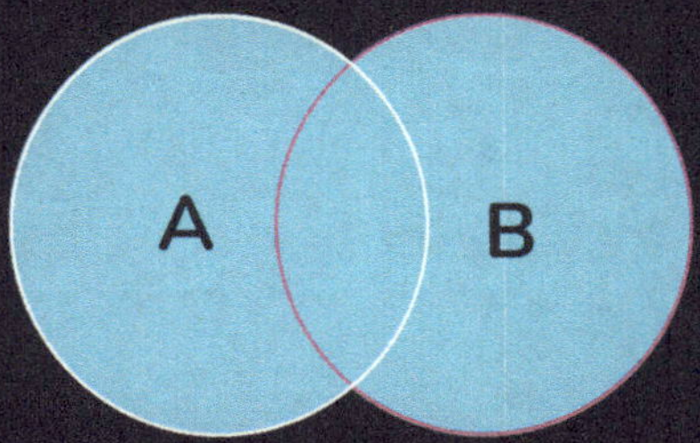

All values in Table A & Table B will be present. When the tables have matching keys, data will be shown in every column. There will only be blanks where the keys do not match.

INNER JOINS

An **INNER JOIN** selects only rows where the join keys match from both tables. Simply stated, only values that are present in both tables are retained.

The figure below illustrates a Venn diagrams for an *INNER JOIN* between Table A and Table B. Notice the area highlighted in blue is where data exists for the specified table. Each join type will be described in-depth on the following pages.

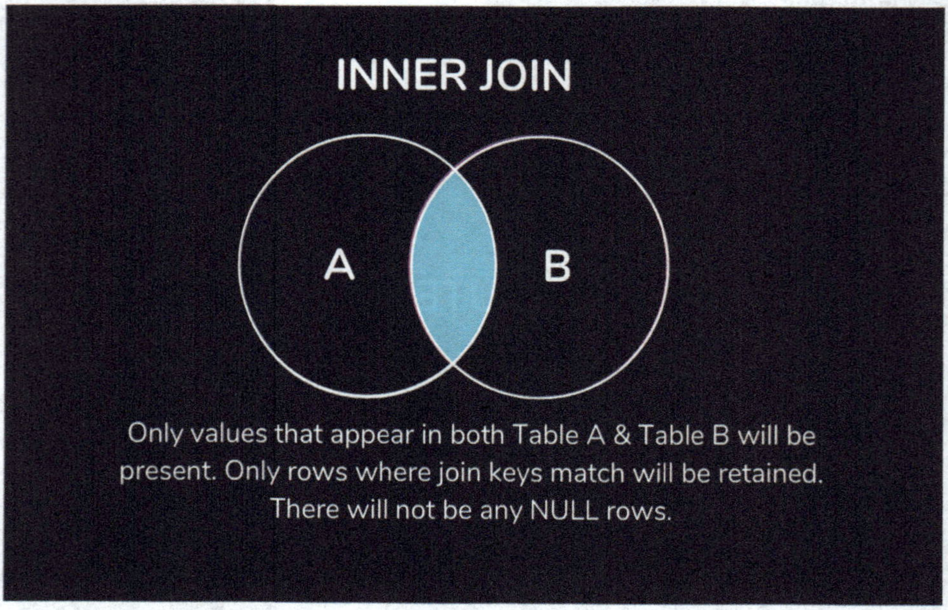

ENVISIONING TABLE JOINS

Envisioning join table output without visual representations can be challenging. For the next few cases, we will use the **STUDENTS** table that is comprised of student enrollment information, and the **GRADES** table that consists of student course grades from the **DATAGURU** database. The Venn diagram below lays out the position of each table (i.e. left or right) for the purpose of the illustrations on the next few pages.

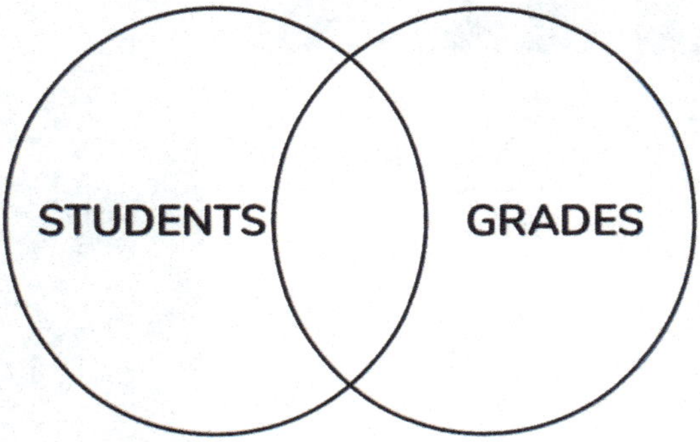

Lets say that you want to see *all enrolled students* and their current grades. A *LEFT JOIN* will allow the **STUDENTS** table to be the master table with *all enrolled student* information retained, but grades will display for only those students that have a matching key. Both, the **STUDENTS** table and the **GRADES** table, have a **StudentID** row which is used as the identifier or join key. A student enrolls in school once, but that same student can have multiple grades. This is considered a one-to-many table relationship.

LEFT JOIN EXAMPLE

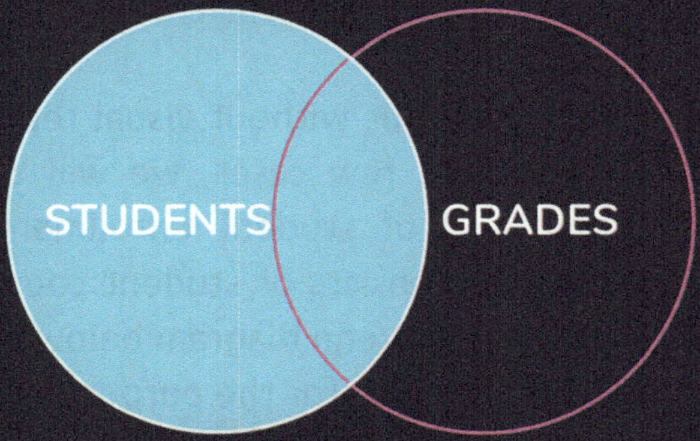

STUDENTS		
StudentID	FirstName	LastName
0000003	Eleonora	Zuniga
0000004	Alyce	Warren
0000005	Nobuko	Moran
0000006	Yee	Esparza
0000007	Chantell	Owens

GRADES		
StudentID	Course	Score
0000001	MATH 120	95
0000002	ENGR 385	67
0000004	ENGR 385	89
0000006	STAT 012	89
0000007	HIST 396	63
0000007	CHEM 281	62

WRITING JOIN QUERIES

When writing a join query, after the *SELECT* statement, the user will indicate the table name and column name(s) from which the data will be retrieved. In the example on the following page, both tables have a column with the same name (i.e. **StudentID**), consquently, the user must clearly signify which table is being addressed. This can be accomplished by referencing the name of the preferred table (typically the master), followed by a period, and then the column name as shown on the next page.

The join type (i.e. left, right, etc.) is stated after the *FROM* clause. In the *FROM* clause, the user indicates the table name (ie, Students), followed by the join type, and lastly the other table name (ie, Grades). Since the **STUDENTS** table is on the left side of the Venn diagram in this example, it will be on the left side of the specified join type.

Finally, the computer must be notified of which keys to join the tables. After the join statement, the key join condition is given. In this example, the join occurs when the **StudentID** is the same in the **STUDENTS** and **GRADES** tables. To depict this condition the *ON* keyword is given, followed by the table name and the join key condition. Again, notice the join condition contains key names which are identical in both tables. In this case, the developer must specify the table name, followed by a period, and then the join key.

LEFT JOIN EXAMPLE WITH HIGHLIGHTS

```
SELECT
    STUDENTS.StudentID,
    FirstName,
    LastName,
    Course,
    Score
FROM STUDENTS
LEFT JOIN GRADES
    ON STUDENTS.StudentID =
      GRADES.StudentID;
```

Left Table

Right Table

Join Condition

Join Type

Join Keys

StudentID	FirstName	LastName	Course	Score
0000004	Alyce	Warren	ENGR 385	89
0000006	Yee	Esparza	STAT 012	89
0000007	Chantell	Owens	HIST 396	63
0000007	Chantell	Owens	CHEM 281	62
0000003	Eleonora	Zuniga	NULL	NULL
0000005	Nobuko	Moran	NULL	NULL

RIGHT JOIN EXAMPLE

When performing a *RIGHT JOIN*, the **GRADES** table will become the master table. All of the grades information will be shown, but only first and last names for students with a matching key will be shown.

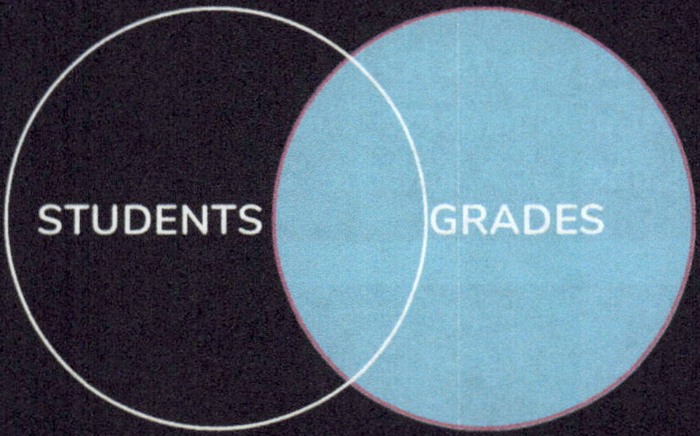

GRADES		
StudentID	Course	Score
0000001	MATH 120	95
0000002	ENGR 385	67
0000004	ENGR 385	89
0000006	STAT 012	89
0000007	HIST 396	63
0000007	CHEM 281	62

STUDENTS		
StudentID	FirstName	LastName
0000003	Eleonora	Zuniga
0000004	Alyce	Warren
0000005	Nobuko	Moran
0000006	Yee	Esparza
0000007	Chantell	Owens

RIGHT JOIN EXAMPLE

GURU TIP

Repeatedly typing the same table name to identify which table to extract data from can be tedious. Instead you can use an alias to shorten the table names as shown below. In the **From** Statment , use the **AS** keyword to alias the tables to assign a shorter unique name. The **STUDENTS** table has been reassigned as A, and the **GRADES** table as B.

```
SELECT
    B.StudentID,
    FirstName,
    LastName,
    Course,
    Score
FROM STUDENTS AS A
RIGHT JOIN GRADES AS B
    ON A.StudentID = B.StudentID;
```

StudentID	FirstName	LastName	Course	Score
0000004	Alyce	Warren	ENGR 385	89
0000006	Yee	Esparza	STAT 012	89
0000007	Chantell	Owens	HIST 396	63
0000007	Chantell	Owens	CHEM 281	62
0000001	NULL	NULL	MATH 120	95
0000002	NULL	NULL	ENGR 385	67

INNER JOIN EXAMPLE

When performing an *INNER JOIN* both tables are considered master tables because all of their information will be retained. Information will be returned if keys match in both tables. Therefore, only students who have grades and grades that correspond with a student will be returned in this example query.

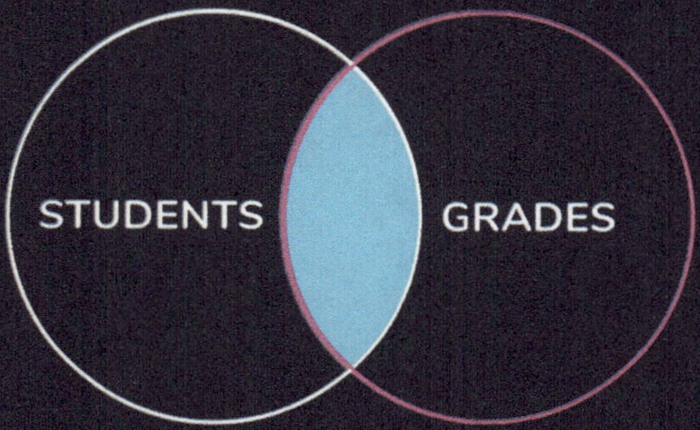

GRADES

StudentID	Course	Score
0000001	MATH 120	95
0000002	ENGR 385	67
0000004	ENGR 385	89
0000006	STAT 012	89
0000007	HIST 396	63
0000007	CHEM 281	62

STUDENTS

StudentID	FirstName	LastName
0000003	Eleonora	Zuniga
0000004	Alyce	Warren
0000005	Nobuko	Moran
0000006	Yee	Esparza
0000007	Chantell	Owens

INNER JOIN EXAMPLE

```
SELECT
    A.StudentID,
    FirstName,
    LastName,
    Course,
    Score
FROM STUDENTS AS A
INNER JOIN GRADES AS B
        ON A.StudentID = B.StudentID;
```

StudentID	FirstName	LastName	Course	Score
0000004	Alyce	Warren	ENGR 385	89
0000006	Yee	Esparza	STAT 012	89
0000007	Chantell	Owens	HIST 396	63
0000007	Chantell	Owens	CHEM 281	62

```
/* Join two tables together using one or
more join conditions */
SELECT
    Column-1
    <,...Column-n>

FROM TABLE-1
    <LEFT, RIGHT, INNER> JOIN TABLE-2

ON condition-1
    <...condition-n>;
```

JOINING DATA

ON THE RIGHT TRACK

Universities and Colleges often have a Career Services office to assist students with getting established on the right track for academic and career success. Some colleges recommend that their students have a career path planned before scheduling classes. In the following exercises you will be assessing data to gain knowledge on students and their respective career track statuses.

EXERCISE 6.1

For this lesson, you will use the **STUDENTS** table, which contains a roster of students currently enrolled at the university, and the **CAREER** table, which lists the various career tracks selected by each student. Each table has a column labeled **StudentID** which uniquely identifies the student and will be used as the join key.

In this task, you will build a report for the Career Services Center to view only the students that have enrolled and designated a career path. The table should only include rows in which the **StudentID** key matches in both tables. *You should not have any blank data values in the final output table.*

First, draw a Venn Diagram to produces the join described in this task. Next, use the pseudocode below to create and run the join query depicted in your Venn Diagram.

```
/*
   Use the STUDENTS and the CAREER tables
   The join type used should only show students who
       exist in both the STUDENTS and CAREER tables
     Select the FirstName, LastName, Track, and Year from the tables
     Use the StudentID, which exists in each table, as the join key
      The join key condition is when both StudentID columns are
   equal
*/
```

EXERCISE 6.2

In this task, you will help the Career Services Center identify students that have enrolled , but have not designated a career track.

Using the **STUDENTS** and **CAREER** tables, draw a *LEFT JOIN* Venn Diagram which represents all students who enrolled from the **STUDENTS** table, along with their career track whether one exists or not in the **CAREER** table.

Next, use the pseudocode below to help you create and run the join query depicted in your Venn Diagram. Note you will add a *WHERE* filter to obtain only registered students who haven't chosen a career track.

```
/*
  Use the STUDENTS and the CAREER tables
  Create a LEFT JOIN which shows all students who registered
     (in STUDENTS table) along with their career track whether
     it exists or not (in CAREER table)
  Select the FirstName, LastName, Track, and Year from the tables
  Use the StudentID in each table as the join key
  The join key condition is when both StudentID columns are equal
  Use a WHERE statement to find where the career Track is
     NULL to see which students haven't chosen a career track
*/
```

EXERCISE 6.3

The purpose of this exercise is to show similarity in the *LEFT JOIN* and *RIGHT JOIN* statements. The results from the last exercise can also be obtained using a RIGHT join.

Using the Venn diagram from the last exercise, switch the table names so that the table on the left is on the right and vice versa. Use the pseudocode below to create and run the join query depicted in your Venn Diagram with the addition of the *WHERE* filter. The final table should contain enrolled students who have not chosen a career track.

```
/*
   Use the CAREER and the STUDENTS tables
   Create a LEFT JOIN from the STUDENTS table to
        the CAREER table
   Select the FirstName, LastName, Track, and Year from the tables
   Use the StudentID in each table as the join key
   The join key condition is when both StudentID columns are equal
   Use a WHERE statement to find records where the career Track is
        NULL to identify students that have not chosen a career track
*/
```

EXERCISE 6.4

When working as a data professional, sometimes odd situations can occur with data. In this task, help the Career Services Center find students that have chosen a career track, but are not currently enrolled at the university. This will help the Career Services Center track down students whose enrollment information which may have been accidentally lost or deleted from the **STUDENTS** dataset, but are showing up in the **CAREER** dataset.

Using the **CAREER** and **STUDENTS** table, draw a *LEFT JOIN* Venn Diagram which represents all students who chose a career track from the **CAREER** table, along with their registration information from the **STUDENTS** table. Next, use the pseudocode below to assist you in creating and running the join query depicted in your Venn Diagram. Note you will add a *WHERE* filter to obtain only those students with a career track who have not registered.

```
/*
  Use the STUDENTS and the CAREER tables
  Create a LEFT JOIN which shows all students who chose a
     career track (in CAREER table) along with their
     registration information registered whether it exists or
     not (in STUDENTS table)
  Select the FirstName, LastName, Track, and Year from the tables
  Use the StudentID in each table as the join key
  The join key condition is when both StudentID columns are equal
  Use a WHERE statement to find where the FirstName is
     NULL to see which students haven't registered
*/
```

EXERCISE 6.5

Similar to exercise 6.3, this exercise will also demonstrate that results from the *LEFT JOIN* can be obtained using a *RIGHT JOIN*. This is another computer programming scenario in which the same answer can be reached with multiple different solutions.

Using the Venn diagram from the last exercise switch the table names where the table on the left is now on the right and vice versa. Use the pseudocode below to assist in creating and running the join query depicted in your Venn Diagram with the addition of the *WHERE* filter to obtain only registered students who have not selected a career track.

```
/*
   Use the STUDENTS and the CAREER tables
   Create a RIGHT JOIN which shows all students who chose a
      career track (in CAREER table) along with their
      registration information registered whether it exists or
      not (in STUDENTS table)
   Select the FirstName, LastName, Track, and Year from the tables
   Use the StudentID in each table as the join key
   The join key condition is when both StudentID columns are equal
   Use a WHERE statement to find where the FirstName is
      NULL to see which students haven't registered
*/
```

ANSWER 6.1

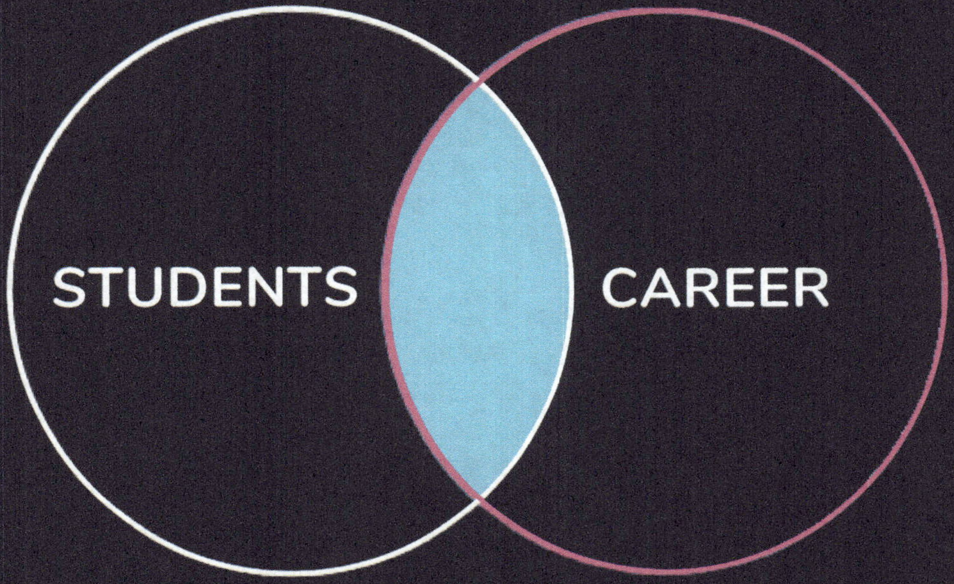

ANSWER 6.1

MySQL 8.0 Command Line Client

```
mysql> SELECT FirstName,
    -> LastName,
    -> Track,
    -> Year
    -> FROM STUDENTS
    -> INNER JOIN CAREER
    -> ON Students.StudentID = Career.StudentID;
+-----------+------------+--------------------+------+
| FirstName | LastName   | Track              | Year |
+-----------+------------+--------------------+------+
| Eleonora  | Zuniga     | Chemical Engineer  | 2020 |
| Yee       | Esparza    | Chemical Engineer  | 2020 |
| Chantell  | Owens      | Project Manager    | 2020 |
| Gisela    | Howell     | Data Scientist     | 2020 |
| Flora     | Singh      | Executive          | 2020 |
| Alfreda   | Benson     | Project Manager    | 2020 |
| Jene      | Whitehead  | Software Developer | 2020 |
| Roxana    | Burke      | Project Manager    | 2020 |
| Verena    | Nolan      | Executive          | 2020 |
| Yuko      | Monroe     | Software Developer | 2020 |
```

ANSWER 6.2

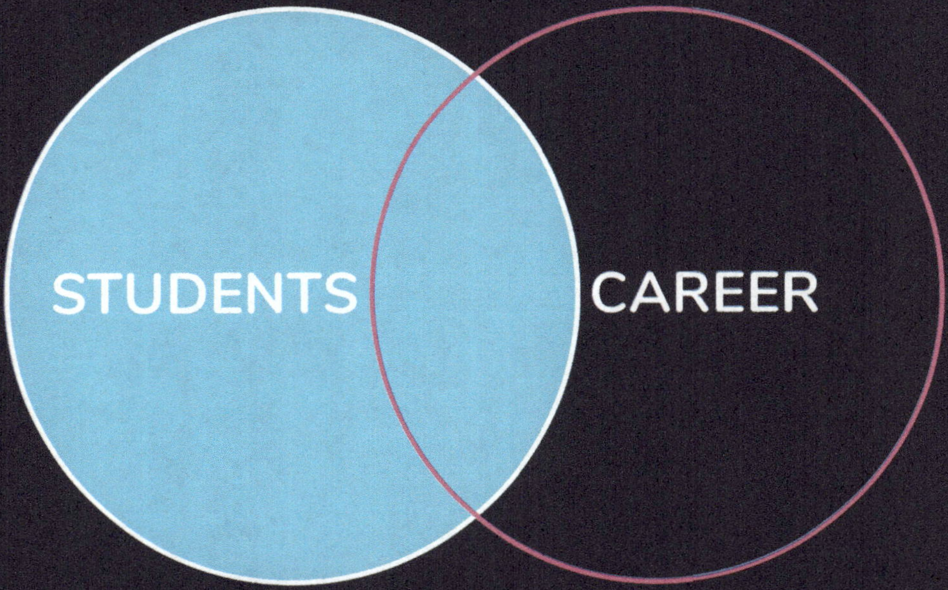

ANSWER 6.2

MySQL 8.0 Command Line Client

```
mysql> SELECT A. StudentID,
    -> FirstName,
    -> LastName,
    -> Track,
    -> Year
    -> FROM STUDENTS AS A
    -> LEFT JOIN CAREER AS B
    -> ON A.StudentID = B.StudentID
    -> WHERE Track IS NULL;
+-----------+------------+-----------+-------+------+
| StudentID | FirstName  | LastName  | Track | Year |
+-----------+------------+-----------+-------+------+
| 0000001   | Mariella   | Figueroa  | NULL  | NULL |
| 0000002   | Cathrine   | Spencer   | NULL  | NULL |
| 0000004   | Alyce      | Warren    | NULL  | NULL |
| 0000005   | Nobuko     | Moran     | NULL  | NULL |
| 0000008   | Vivian     | Humphrey  | NULL  | NULL |
| 0000009   | Branda     | Galvan    | NULL  | NULL |
| 0000011   | Mathilda   | Gordon    | NULL  | NULL |
```

ANSWER 6.3

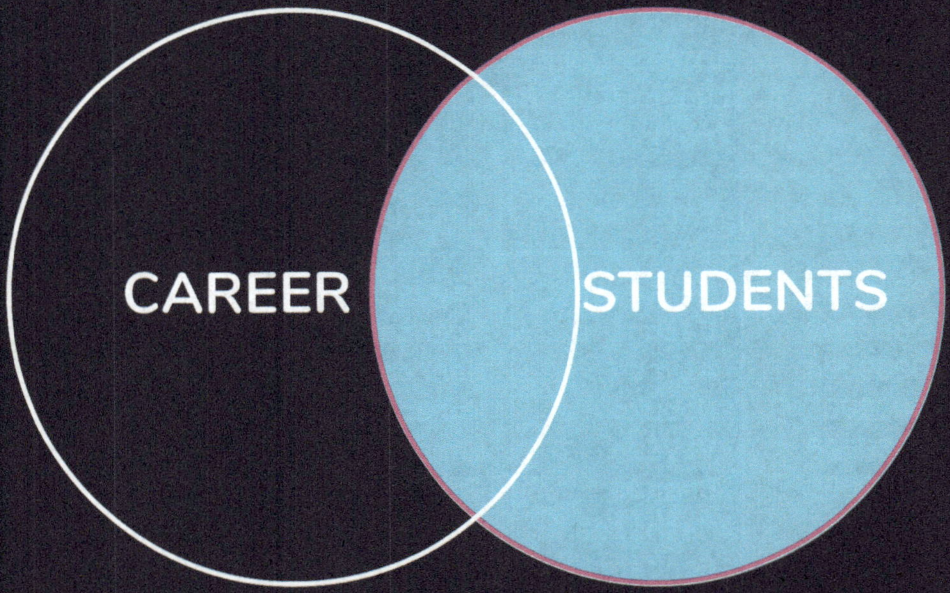

ANSWER 6.3

MySQL 8.0 Command Line Client

```
mysql> SELECT B. StudentID,
    -> FirstName,
    -> LastName,
    -> Track,
    -> Year
    -> FROM CAREER AS A
    -> RIGHT JOIN STUDENTS AS B
    -> ON A.StudentID = B.StudentID
    -> WHERE Track IS NULL;
+-----------+-------------+-----------+-------+------+
| StudentID | FirstName   | LastName  | Track | Year |
+-----------+-------------+-----------+-------+------+
| 0000001   | Mariella    | Figueroa  | NULL  | NULL |
| 0000002   | Cathrine    | Spencer   | NULL  | NULL |
| 0000004   | Alyce       | Warren    | NULL  | NULL |
| 0000005   | Nobuko      | Moran     | NULL  | NULL |
| 0000008   | Vivian      | Humphrey  | NULL  | NULL |
| 0000009   | Branda      | Galvan    | NULL  | NULL |
| 0000011   | Mathilda    | Gordon    | NULL  | NULL |
| 0000012   | Conception  | Carpenter | NULL  | NULL |
| 0000014   | Dixie       | Bullock   | NULL  | NULL |
```

ANSWER 6.4

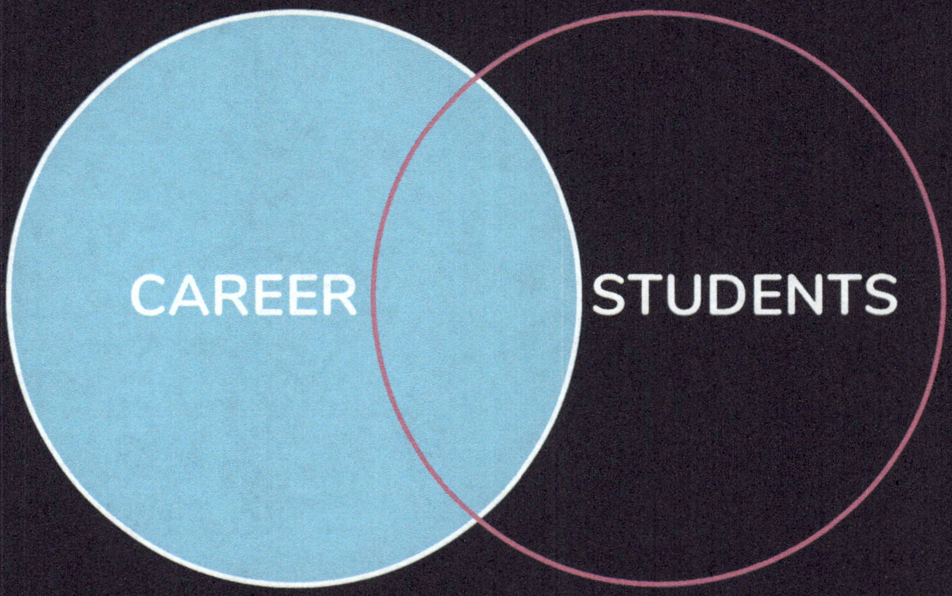

ANSWER 6.4

MySQL 8.0 Command Line Client

```
mysql> SELECT A. StudentID,
    -> FirstName,
    -> LastName,
    -> Track,
    -> Year
    -> FROM CAREER AS A
    -> LEFT JOIN STUDENTS AS B
    -> ON A.StudentID = B.StudentID
    -> WHERE FirstName is NULL;
+-----------+-----------+----------+------------------------+------+
| StudentID | FirstName | LastName | Track                  | Year |
+-----------+-----------+----------+------------------------+------+
| 0000107   | NULL      | NULL     | Cyber Security Analyst | 2021 |
| 0000108   | NULL      | NULL     | Cyber Security Analyst | 2021 |
| 0000109   | NULL      | NULL     | Cyber Security Analyst | 2021 |
| 0000110   | NULL      | NULL     | Cyber Security Analyst | 2021 |
| 0000111   | NULL      | NULL     | Cyber Security Analyst | 2021 |
| 0000112   | NULL      | NULL     | Cyber Security Analyst | 2021 |
| 0000113   | NULL      | NULL     | Cyber Security Analyst | 2021 |
| 0000114   | NULL      | NULL     | Cyber Security Analyst | 2021 |
| 0000115   | NULL      | NULL     | Cyber Security Analyst | 2021 |
| 0000116   | NULL      | NULL     | Cyber Security Analyst | 2021 |
| 0000117   | NULL      | NULL     | Cyber Security Analyst | 2021 |
| 0000118   | NULL      | NULL     | Cyber Security Analyst | 2021 |
| 0000119   | NULL      | NULL     | Cyber Security Analyst | 2021 |
```

ANSWER 6.5

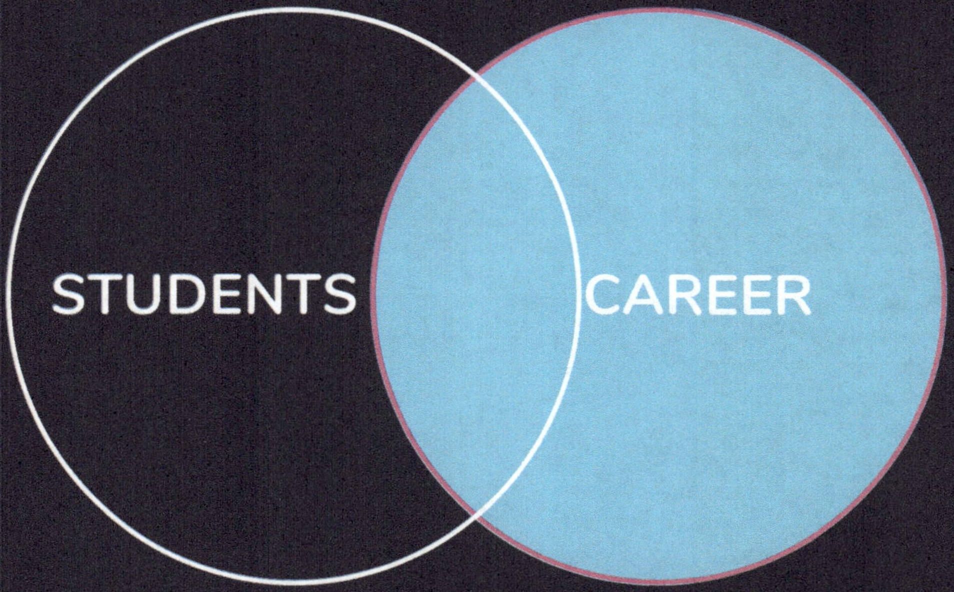

ANSWER 6.5

MySQL 8.0 Command Line Client

```
mysql> SELECT B.StudentID,
    -> FirstName,
    -> LastName,
    -> Track,
    -> Year
    -> FROM STUDENTS AS A
    -> RIGHT JOIN CAREER AS B
    -> ON A.StudentID = B.StudentID
    -> WHERE FirstName is NULL;
+-----------+-----------+----------+-------------------------+------+
| StudentID | FirstName | LastName | Track                   | Year |
+-----------+-----------+----------+-------------------------+------+
| 0000107   | NULL      | NULL     | Cyber Security Analyst  | 2021 |
| 0000108   | NULL      | NULL     | Cyber Security Analyst  | 2021 |
| 0000109   | NULL      | NULL     | Cyber Security Analyst  | 2021 |
| 0000110   | NULL      | NULL     | Cyber Security Analyst  | 2021 |
| 0000111   | NULL      | NULL     | Cyber Security Analyst  | 2021 |
| 0000112   | NULL      | NULL     | Cyber Security Analyst  | 2021 |
| 0000113   | NULL      | NULL     | Cyber Security Analyst  | 2021 |
| 0000114   | NULL      | NULL     | Cyber Security Analyst  | 2021 |
| 0000115   | NULL      | NULL     | Cyber Security Analyst  | 2021 |
| 0000116   | NULL      | NULL     | Cyber Security Analyst  | 2021 |
| 0000117   | NULL      | NULL     | Cyber Security Analyst  | 2021 |
| 0000118   | NULL      | NULL     | Cyber Security Analyst  | 2021 |
| 0000119   | NULL      | NULL     | Cyber Security Analyst  | 2021 |
```

SUMMARY

In this chapter you learned:

- **JOIN** statements allow users to horizontally combine tables by row based on one or more matching columns or criteria.

- **Cardinality** is a term used to refer to the uniqueness of data values and to understand the relationship between two or more datasets in a RDBMS. The different relationship types are referred to as: **one-to-one**, **one-to-many**, and **many-to-many**.

- Developers utilize one or more **keys** or row identifiers to relate one or more tables.

- A **primary key** represents a unique identifier for each row. Each identifier must be different and no two rows will have the same identifier(i.e. high cardinality).

- **Secondary keys** are also used as identifiers, but are not required to be unique and may be shared among rows (i.e. low cardinality).

- In a **LEFT JOIN** the left table is the master table in the join and in a RIGHT JOIN the right table is the master table in the join

- In a **FULL OUTER JOIN** both tables are equally important to the join. The join returns unmatched rows from both tables.

- An **INNER JOIN** selects only the rows where the join keys match in both datasets.

“

Doing the best
at this moment...

Puts you in the
best place for
the next moment

OPRAH WINFREY

DAY 7

In this chapter, you will:

- Learn about Big Data challenges and a technique used to rectify them
- Merge tables vertically using Unions
- Use vertical merges and horizontal joins simultaneously

DAY 7 OVERVIEW

Welcome to Day 7!

Only a few more pages to go and you've achieved Data Guru status!

In the previous chapter you learned about joins. Joins are one of the most complex concepts in data management. Today you will gain an understanding of two additional methods used to merge tables in SQL. The final tasks that remain to complete this book are shown below:

1) Plan and diagram table joins
2) Stack or combine tables vertically
3) Join tables vertically using an Outer Join

In this lesson, you will observe what happens when you add a large number of new rows to an existing database. This is a common occurrence in database management since data is continuously collected in the digital world.

GURU TIP

Since you've made it this far in the book, you probably know that a big part of being a Data Guru is keeping your skillset sharp. **Never forget that practice makes perfect.** For more practice, tips, tricks, and training , head to guru.dn2g.com

DATA CHUNKING

As stated in the introductory chapter, the world is producing a large amount of information and along with this increase comes challenges with Big Data . **Big Data** is a term that represents four challenges organizations face when dealing with information. They include dilemmas with volume, velocity, variety, and veracity. The "Four V's of Big Data" are defined as follows:

Volume - Data is large in size;
Velocity - Data is processed at a rapid rate;
Variety - Data has many different forms and sources;
Veracity - Data that is dirty or has low quality.

Frequently, database users must access large volumes of data at rapid velocities. Storing, accessing, analyzing, and transferring large amounts of data can be computationally stressful on databases resulting in poor data quality, slow speeds, and overall decreased productivity. One method that helps with faster data transport, storage, and access is to split the data into reasonable sized pieces called **chunks**.

Data chunking is a useful way to quickly share information across servers. A single large chunk of data can take a long time to transmit from the source to the target server, but multiple small data files can transmit faster and in parallel (i.e. simultaneously). Once the data reaches its target destination, it will likely need to be reassembled for use in analysis. This chapter describes how to reassemble data.

GURU TIP

When chunking data to export to a file and transport, use the *WHERE* filter to separate data into different tables. Ensure the tables are labeled so you can identify each of them.

DATA CHUNK REASSEMBLY DIAGRAM

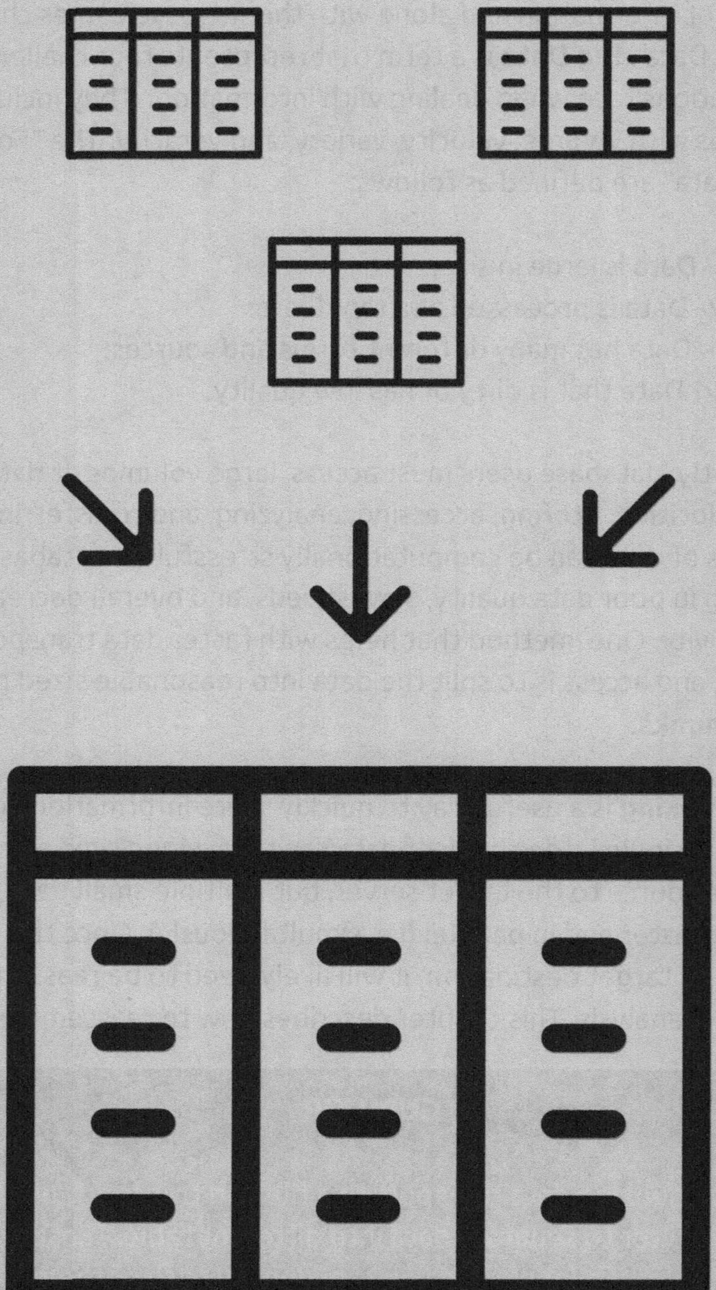

UNION & UNION ALL

The previous chapter covered how to join two tables horizontally, but there are also situations in which vertical joins, or appending tables, are beneficial. Appending tables or stacking one table on top of another is called a **UNION** join in SQL.

A common use case for a *UNION* is when a developer receives a dataset in batches. When moving large datasets from one server to another, a common practice is to separate the dataset into sizable chunks. By transmitting smaller batches of data, a computer can send and retrieve data faster. Once the chunked data reaches its destination, it may be reassembled using a *UNION* statement.

When performing a union of two or more datasets, they must have the same column names and corresponding data types must exist in all tables. These rules exist so that users can't combine two mismatch tables and compromise the integrity of either dataset. Note: When rows containing the same values exist in two or more tables, the *UNION* command automatically removes the other duplicate rows leaving only one unique row. However, if the user desires to keep duplicate rows the command **UNION ALL** will keep all values including duplicates.

UNION & UNION ALL EXAMPLE

STUDENTS		
StudentID	**FirstName**	**LastName**
0000001	Mariella	Figueroa
0000002	Cathrine	Spencer
0000003	Eleonora	Zuniga
0000004	Alyce	Warren
0000005	Nobuko	Moran

↕

NEWSTUDENTS		
StudentID	**FirstName**	**LastName**
0000001	Mariella	Figueroa
0000002	Cathrine	Spencer
0000004	Alyce	Warren
0000006	Yee	Esparza
0000007	Chantell	Owens

In the example above, there are two tables **STUDENTS** and **NEWSTUDENTS** with the same column names and data types. There are a few rows which contain exactly the same data in both tables (i.e. where **StudentID** is "0000001", "0000002", and "0000004"). View the next two pages to see how these duplicate rows are handled using *UNION* versus *UNION ALL*.

```
SELECT StudentID,
    FirstName,
    LastName
FROM STUDENTS
    UNION
SELECT StudentID,
    FirstName,
    LastName
FROM NEWSTUDENTS;
```

StudentID	FirstName	LastName
0000001	Mariella	Figueroa
0000002	Cathrine	Spencer
0000003	Eleonora	Zuniga
0000004	Alyce	Warren
0000005	Nobuko	Moran
0000006	Yee	Esparza
0000007	Chantell	Owens

Notice when using the *UNION* statement how only one of the duplicate rows is kept where **StudentID** is "0000001", "0000002", and "0000004".

UNION ALL

```
SELECT StudentID,
       FirstName,
       LastName
FROM STUDENTS
       UNION ALL
SELECT StudentID,
       FirstName,
       LastName
FROM NEWSTUDENTS;
```

StudentID	FirstName	LastName
0000001	Mariella	Figueroa
0000002	Cathrine	Spencer
0000003	Eleonora	Zuniga
0000004	Alyce	Warren
0000005	Nobuko	Moran
0000001	Mariella	Figueroa
0000002	Cathrine	Spencer
0000004	Alyce	Warren
0000006	Yee	Esparza
0000007	Chantell	Owens

Conversely, when using the *UNION ALL* statement, notice there are duplicate instances where **StudentID** is equal to "0000001", "0000002", and "0000004".

FULL OUTER JOIN

Full Outer Join will be the final topic covered in this book. There is definitely a reason this topic was saved for last.

A **FULL OUTER JOIN** occurs when the user joins two tables horizontally, but needs to retain data from both tables even if the two tables do not match. Therefore, both tables are equally important and both assume the Master role.

A *FULL OUTER JOIN* returns all columns specified by the user from both tables whether the keys match or not. When the keys match,, the output will display data in the row. When the keys do not match, blank areas will display in the output table. However, you will retain all data from both tables.

The syntax to perform a *FULL OUTER JOIN* in MySQL is a combination of the *LEFT JOIN*, *RIGHT JOIN*, and *UNION* statement. Now you are probably starting to understand why this topic was saved for last. The next few pages provide more detail on how to write a *FULL OUTER JOIN* query.

GURU TIP

In MySQL, implementing a *FULL OUTER JOIN* command requires numerous steps. However, this is not the case for all SQL software providers. Some SQL software have a *FULL OUTER JOIN* command although MySQL does not.

FULL OUTER JOIN VENN DIAGRAM

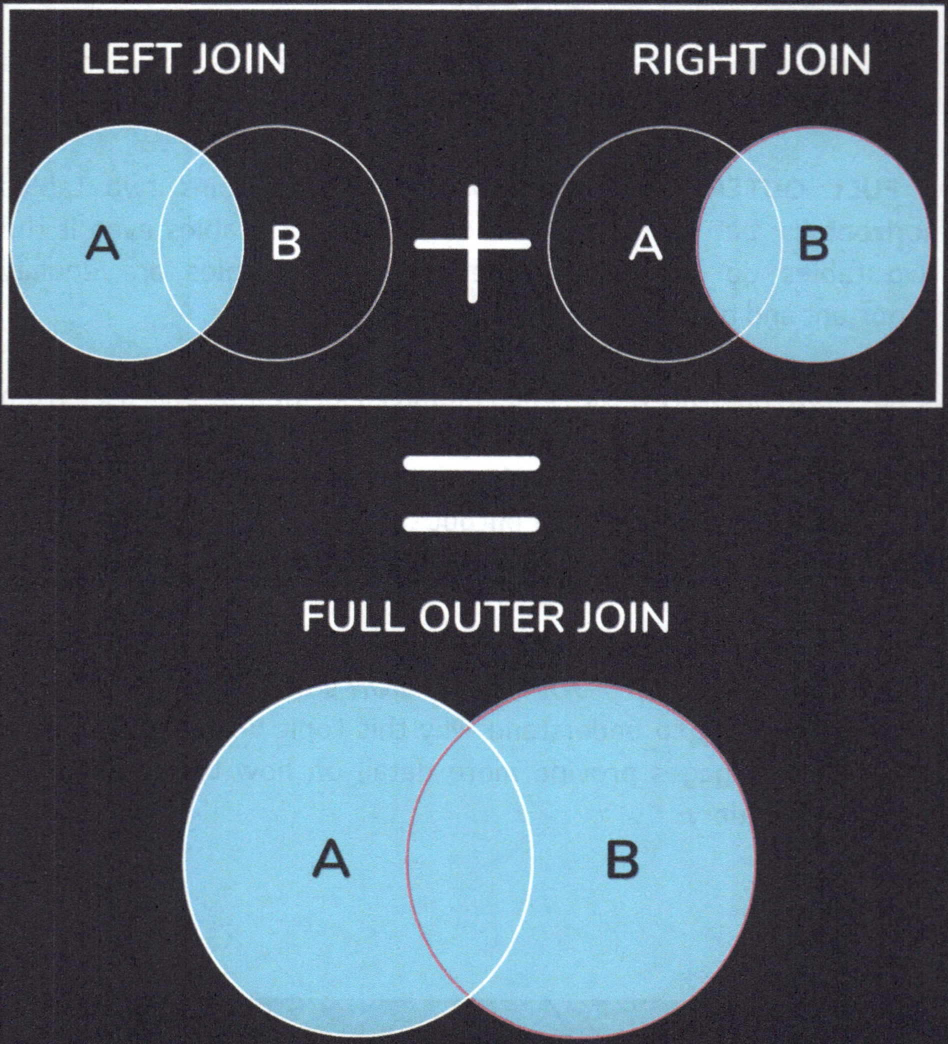

In the depiction above, notice the left join of Table A and B unioned with the *RIGHT JOIN* of Table A and B equates to a *FULL OUTER JOIN.* A more detailed example along with table output is shown on the following pages.

WRITING FULL OUTER JOIN QUERIES

The next few pages demonstrate the steps required to perform a *FULL OUTER JOIN* using data extracted from the **STUDENTS** and **GRADES** tables located in the **DATAGURU** database. This join type allows the developer to combine all data in both tables, along with any rows that match on a given join condition.

The first step when writing a *FULL OUTER JOIN* query, is to perform a *LEFT JOIN*. The next step is to perform a *RIGHT JOIN*. The join keys, conditions, and query logic must be the same aside from the type of join being used (i.e. left join or right join). Once the two joins are performed they are unionied to complete the *FULL OUTER JOIN*.

In the examples, both tables have a **StudentID** column, so the user much specify which table to pull this column from when listing this column name in the *LEFT JOIN* and *RIGHT JOIN* queries. Do this by referencing the name of the preferred table (typically the master), followed by a period, and the column name as shown on the following pages.

STEP 1: TYPE LEFT JOIN

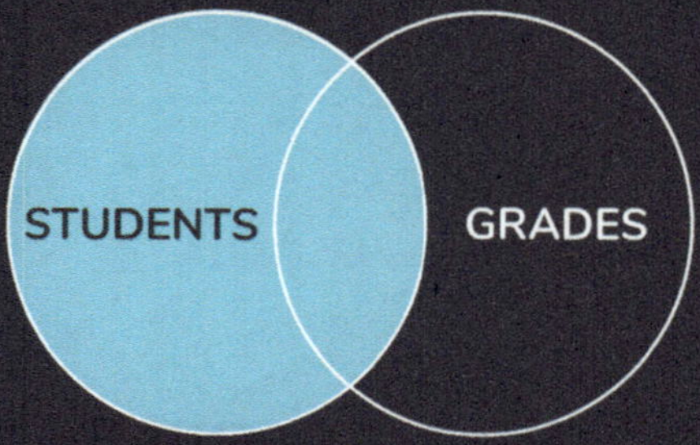


```
SELECT
    STUDENTS.StudentID,
    FirstName,
    LastName,
    Course,
    Score
FROM STUDENTS
    LEFT JOIN GRADES ON
    STUDENTS.StudentID =
    GRADES.StudentID
```

StudentID	FirstName	LastName	Course	Score
0000004	Alyce	Warren	ENGR 385	89
0000006	Yee	Esparza	STAT 012	89
0000007	Chantell	Owens	HIST 396	63
0000007	Chantell	Owens	CHEM 281	62
0000003	Eleonora	Zuniga	NULL	NULL
0000005	Nobuko	Moran	NULL	NULL

STEP 2: TYPE RIGHT JOIN

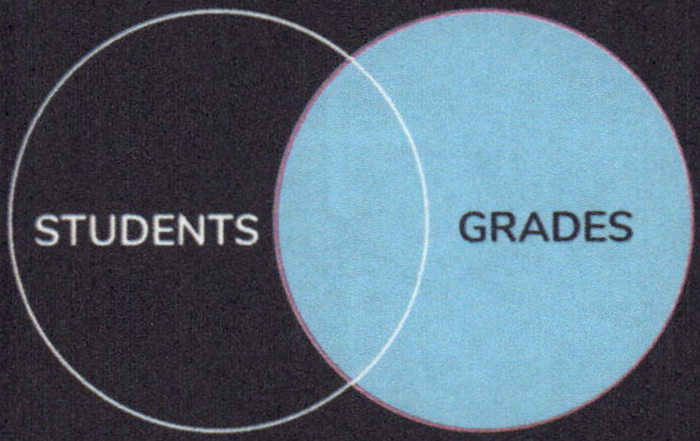


```
SELECT
    GRADES.StudentID,
    FirstName,
    LastName,
    Course,
    Score
FROM STUDENTS
    RIGHT JOIN GRADES ON
    STUDENTS.StudentID =
    GRADES.StudentID;
```

StudentID	FirstName	LastName	Course	Score
0000004	Alyce	Warren	ENGR 385	89
0000006	Yee	Esparza	STAT 012	89
0000007	Chantell	Owens	HIST 396	63
0000007	Chantell	Owens	CHEM 281	62
0000001	NULL	NULL	MATH 120	95
0000002	NULL	NULL	ENGR 385	67

STEP 3. UNION LEFT AND RIGHT JOINS

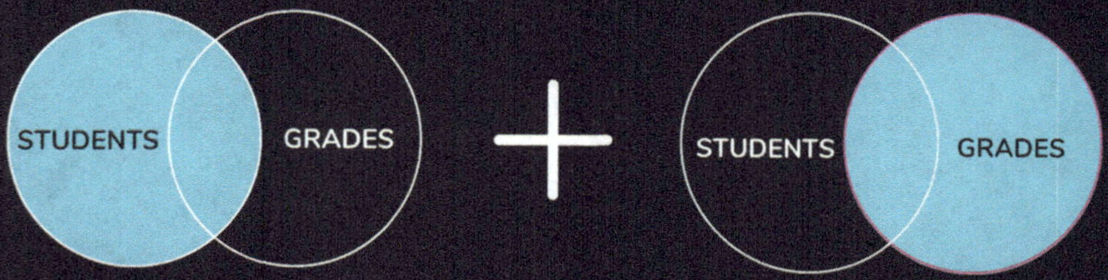


```
SELECT
    STUDENTS.StudentID,
    FirstName,
    LastName,
    Course,
    Score
FROM STUDENTS
LEFT JOIN GRADES
    ON STUDENTS.StudentID =
    GRADES.StudentID
    UNION
SELECT
    GRADES.StudentID,
    FirstName,
    LastName,
    Course,
    Score
FROM STUDENTS
RIGHT JOIN GRADES
    ON STUDENTS.StudentID =
    GRADES.StudentID;
```

FULL OUTER JOIN EXAMPLE OUTPUT

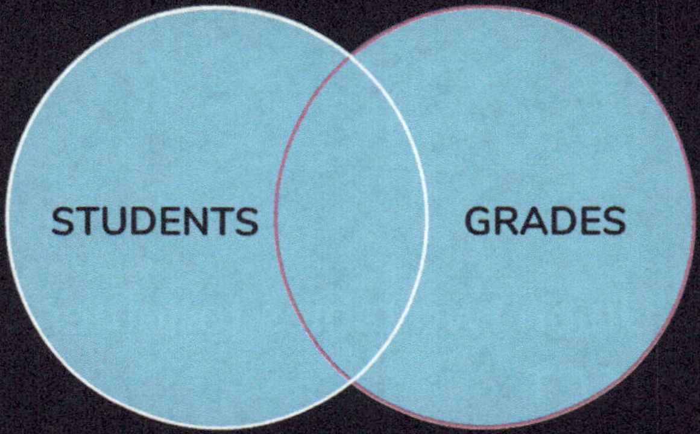

StudentID	FirstName	LastName	Course	Score
0000004	Alyce	Warren	ENGR 385	89
0000006	Yee	Esparza	STAT 012	89
0000007	Chantell	Owens	HIST 396	63
0000007	Chantell	Owens	CHEM 281	62
0000003	Eleonora	Zuniga	NULL	NULL
0000005	Nobuko	Moran	NULL	NULL
0000001	NULL	NULL	MATH 120	95
0000002	NULL	NULL	ENGR 385	67

The output from the *FULL OUTER JOIN* results in a table which contains all data from both the **STUDENTS** and **GRADES** tables. Notice, where the join conditions are not met, there are missing or NULL values for the data elements highlighted. Additionally, the *UNION* statement was used which resin the removal of duplicate rows.

UNION SYNTAX

```
-- Union two tables together
SELECT
    Column-1
    <,...Column-n>

FROM TABLE-1

    UNION <ALL>

SELECT
    Column-1
    <,...Column-n>

FROM TABLE-2;
```

FULL OUTER JOIN SYNTAX

```
-- Union two tables together
SELECT
    Column-1
    <,...Column-n>

FROM TABLE-1
    LEFT JOIN TABLE-2
    ON condition(s)

        UNION <ALL>

SELECT
    Column-1
    <,...Column-n>

FROM TABLE-1
    LEFT JOIN TABLE-2
    ON condition(s);
```

LESSON 7

UNION DATA

NEW EMPLOYEE HIRES

In database management, new data is constantly being streamed and added to databases. It is pertinent to understand how to augment your database since this will be a regular occurrence. For example, organizations are continuously hiring new employees which means systematic changes to tables in their RDBMS. In this lesson, you will manage college faculty data to account for both new hires and existing faculty members.

EXERCISE 7.1

In this lesson, you will combine two tables, one that contains existing faculty at the university (**FACULTY**) and another that is comprised of new faculty hires (**NEWFACULTY**). Use the pseudocode below to create a new table that is a combination of the two tables. Make sure no duplicate rows exist.

```
/*
  Create a table called UNION_FACULTY
  Select all of the columns from the FACULTY table
  UNION them with all the columns from the NEWFACULTY table
      in such a way that duplicate rows are removed
*/
```

EXERCISE 7.2

In the previous exercise, two tables were combined and all duplicates were removed from the newly created table. Create another dataset which combines the **FACULTY** and **NEWFACULTY** datasets. Retain all rows including duplicate rows, if they exist.

```
/*
  Create a table called UNIONALL_FACULTY
  Select all of the columns from the FACULTY table
  UNION ALL with all the columns from the NEWFACULTY table
      to combine tables and keep any duplicate rows
*/
```

EXERCISE 7.3

The next few exercises deconstruct the steps required to implement a *FULL OUTER JOIN* in MySQL. Your overall objective, which is the output of the Outer Join query, is to figure out the following:

- Which courses don't have a faculty member assigned to them?
- Which faculty members don't have courses assigned to them?

For the first step in this task, use the pseudocode below to create a table containing a unique list of courses offered at the university including their corresponding semester and year.

```
/*
  Create a new table called COURSES
  Select the unique values from the Course, Semester, and
     Year columns  located in the GRADES table
*/
```

EXERCISE 7.4

In this exercise, you will use the **FACULTY** table, which consists of the **UNION_FACULTY** currently teaching at the university, along with **COURSES** table that lists the courses offered. Each table has a column labeled Course which is an identifier to be used as the join key.

You have been tasked to assist the school by building a report that displays all faculty at the institution along with the respective courses they are assigned to teach. The table should include all rows from the **UNION_FACULTY** table and may contain blanks for the **COURSES** data if there are no matching Course identifiers.

First, sketch a *LEFT JOIN* Venn Diagram which depicts the scenario described above. Next, use the pseudocode below to assist you in creating and running the join query depicted in your Venn Diagram.

```
/*
  Create a new table called LEFT_OUTER
  Use the UNION_FACULTY(on left) and the COURSES (on
     right) tables
  Create a LEFT JOIN which shows all faculty members (in
     FACULTY table) along with the courses they taught whether
     they taught or not (in COURSES table)
  Select the FirstName, LastName, Course (from FACULTY
     table), Semester, and Year from the tables
  Use the Course in each table as the join key
  The join key condition is when both Course columns are equal
*/
```

EXERCISE 7.5

In this task, you will help the school build a report they can use to view all course information and the faculty assigned to teach them.

Using the **COURSES** and **UNION_FACULTY** table, draw a *RIGHT JOIN* Venn Diagram which represents all courses shown in the **COURSES** table along with their assigned faculty member, whether or not one has been assigned in the **UNION_FACULTY** table. Next, use the pseudocode below to assist you in creating and running the join query depicted in your Venn Diagram.

```
/*
  Create a new table called RIGHT_OUTER
  Use the UNION_FACULTY (on left) and the COURSES (on right)
     tables
  Create a RIGHT JOIN which shows all courses (in COURSES
     table)    along with the faculty assigned to teach the
     course, whether or not a teacher has been assigned
      (in UNION_FACULTY table)
  Select the FirstName, LastName, Course (from COURSES
     table), Semester, and Year from the tables
  Use the Course in each table as the join key
  The join key condition is when both Course columns are equal
*/
```

EXERCISE 7.6

In this exercise, use the queries from Exercises 7.3 and 7.4 to create a *FULL OUTER JOIN* on the **UNION_FACULTY** and **COURSES** table. Using the *LEFT JOIN* and *RIGHT JOIN* queries, *UNION* the statements. The final table which will show which faculty members are missing courses and which courses are missing teachers.

```
/*
  Create a new table called FACULTY_COURSES
  Select all of the columns from the LEFT_OUTER table
  UNION them with all the columns from the RIGHT_OUTER table
      to ensure any duplicate rows are removed
*/
```

ANSWER 7.1

MySQL 8.0 Command Line Client

```
mysql> CREATE TABLE UNION_FACULTY AS
    -> SELECT *
    -> FROM FACULTY
    -> UNION
    -> SELECT *
    -> FROM NEWFACULTY;
Query OK, 36 rows affected (0.27 sec)
Records: 36  Duplicates: 0  Warnings: 0
```

ANSWER 7.2

MySQL 8.0 Command Line Client

```
mysql> CREATE TABLE UNIONALL_FACULTY AS
    -> SELECT *
    -> FROM FACULTY
    -> UNION ALL
    -> SELECT *
    -> FROM NEWFACULTY;
Query OK, 43 rows affected (0.07 sec)
Records: 43  Duplicates: 0  Warnings: 0

mysql>
```

ANSWER 7.3

MySQL 8.0 Command Line Client

```
mysql> CREATE TABLE COURSES AS
    -> SELECT DISTINCT Course,
    -> Semester,
    -> Year
    -> FROM GRADES;
Query OK, 20 rows affected (0.07 sec)
Records: 20  Duplicates: 0  Warnings: 0

mysql>
```

ANSWER 7.4

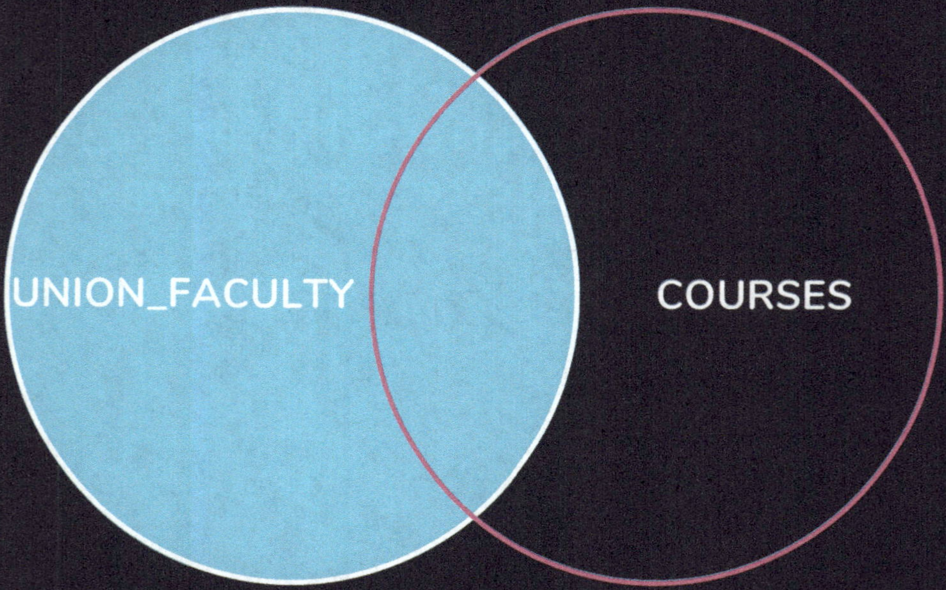

ANSWER 7.4

MySQL 8.0 Command Line Client

```
mysql> CREATE TABLE LEFT_OUTER AS
    -> SELECT FirstName,
    -> LastName,
    -> A.Course,
    -> Semester,
    -> Year
    -> FROM UNION_FACULTY AS A
    -> LEFT JOIN COURSES AS B
    -> ON A.Course=B.Course;
Query OK, 103 rows affected (0.05 sec)
Records: 103  Duplicates: 0  Warnings: 0

mysql>
```

ANSWER 7.5

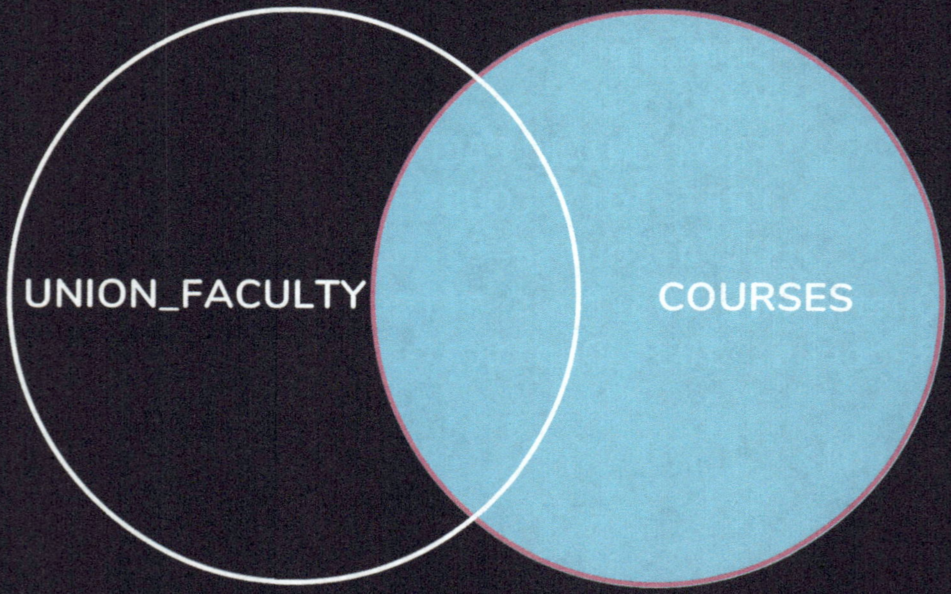

ANSWER 7.5

MySQL 8.0 Command Line Client

```
mysql> CREATE TABLE RIGHT_OUTER AS
    -> SELECT FirstName,
    -> LastName,
    -> B.Course,
    -> Semester,
    -> Year
    -> FROM UNION_FACULTY AS A
    -> RIGHT JOIN COURSES AS B
    -> ON A.Course=B.Course;
Query OK, 103 rows affected (0.07 sec)
Records: 103  Duplicates: 0  Warnings: 0

mysql>
```

ANSWER 7.6

MySQL 8.0 Command Line Client

```
mysql> CREATE TABLE FACULTY_COURSES AS
    -> SELECT * FROM LEFT_OUTER
    -> UNION
    -> SELECT * FROM RIGHT_OUTER;
Query OK, 103 rows affected (0.06 sec)
Records: 103  Duplicates: 0  Warnings:
0

mysql>
```

SUMMARY

In this chapter you learned:

- **Big Data** is a term which represents four issues organizations face when dealing with information including struggles with volume, velocity, variety, and veracity.

- One of the methods to help with faster data transport, storage, and access is to separate the data up into reasonable sized pieces called **chunks**.

- **Data chunking is** a beneficial method to quickly share information across servers.

- Appending tables or stacking one table on top of another is called a **UNION** join in SQL.

- The **UNION** command automatically removes duplicate rows. However, if the user desires to retain duplicate values use the **UNION ALL** command instead.

- A **FULL OUTER JOIN** occurs when the developer needs to join two tables horizontally, but needs to retain data from all tables whether or not there are matches between the two.

"

Vini

Vidi

Vici

YOU
THEE DATA GURU

In the tech industry, you will:

- Go forth and take on Big Data Challenges
- Begin analyzing data like a Fortune 100 Pro
- Keep learning because you can never be too much of an expert

DEFINITIONS

Aggregate Functions - special type of functions which allow the user to look at summary data or statistics

Alter Table - a way to make changes to an existing dataset or table

Big Data - represents volume, velocity, variety, and veracity challenges people face when dealing with complex information

Business Intelligence & Analytics (BI & A) - use of business-related information to strategicially achieve company objectives

Cardinality - refers to the uniqueness of data values in table columns

CASE Statement - statement that allows the computer to check through a list of one or more conditions and complete an action when one of the conditions is true

Cluster - a group of data that has similar characteristics

Clustering - technique in data analysis used to group and detect patterns among groups

Coding - also called computer programming, allows developers to send instructions to a computer to complete a specific set of actions

Columnar Databases - comparable to tables in a relational database except each column is segmented into a separate file and stored

Comma Separated Value (CSV) - a type of text document in which all the data elements are separated or delimited by commas

Command Line Interface (CLI) - terminal in which the user can submit instructions to the computer

DEFINITIONS

Comments - used to create notes in code without interfering with syntax rules

Condition - a statement that can be either true or false

Data - any recorded information which may or may not be used to gain insight

Data Chunking - a useful way to split large data for faster transport and easier data management

Data Imputation - the substitution of missing data values

Data Transformation - the act of manipulating, subsetting, or even substituting data

Database - collection of machine-readable information, or datasets, organized so that it can be easily accessed, managed and updated

Database Management System (DBMS) - collection of integrated services which support database management and control the creation, use and maintenance of a database

Dataset - a group or collection of data

Debugging - the process of finding and removing coding errors.

DISTINCT - SQL keyword used to find unique values

Document Store Databases - a type of data storage in which data is stored in an XML or JSON like document

Filtering - allows the user to extract a portion of data in which specified values meet a certain condition

DEFINITIONS

FROM - SQL keyword used to tell the computer which table to extract data

FULL OUTER JOIN - SQL join type which returns all columns specified by the user from both tables whether the keys match or not

Functions - code which allows the user to perform a routine process or repeated sequence of steps

Graph-Based Databases - data storage type which uses a network-like structure called a graph that is made up of nodes and edges

GROUP BY - SQL statement that allows the user to obtain metrics across categories

INNTER JOIN - SQL join type which returns only rows where the join keys match from both tables

Integers - refers to whole numbers that do not contain decimal points, represent fractional values, or other imaginary numbers.

Key-Value Databases - data storage type which stores information in pairs; the parent object is called the key and the child object is called the value

Keywords - standard words in computer language which are used to complete certain actions or a series of commands

Mathematical Functions - used to perform tasks, manipulate numeric data, represent complex numbers, or to depict mathematical symbols

DEFINITIONS

Non-Relational Databases - also known as NoSQL databases, may or may not store data in a defined structure such as a table, but datasets are not required to relate to one another.

Numeric - data types that are used to store quantifiable values represented by numbers

ORDER BY - SQL keyword that allows the user to sort a dataset based on the values located in one or more columns

Primary Key - A unique identifier for each row in a table

Secondary Key - An identifier for each row in a table, but is not required to be unique

WHERE - SQL keyword used to add filtering logic to database queries

INDEX

INDEX

Made in United States
North Haven, CT
04 November 2026

21730345R00163